Molecular Medical Biochemistry

Molecular medical biochemistry

J. P. LUZIO

University Lecturer in Clinical Biochemistry,
University of Cambridge;
Fellow, St Edmund's College, Cambridge

R. J. THOMPSON

Professor of Clinical Biochemistry,
University of Southampton,
(sometime Fellow and Director of
Medical Studies, Corpus Christi College, Cambridge)

*The right of the
University of Cambridge
to print and sell
all manner of books
was granted by
Henry VIII in 1534.
The University has printed
and published continuously
since 1584.*

CAMBRIDGE UNIVERSITY PRESS

CAMBRIDGE
NEW YORK   PORT CHESTER
MELBOURNE   SYDNEY

CAMBRIDGE UNIVERSITY PRESS
Cambridge, New York, Melbourne, Madrid, Cape Town, Singapore, São Paulo

Cambridge University Press
The Edinburgh Building, Cambridge CB2 8RU, UK

Published in the United States of America by Cambridge University Press, New York

www.cambridge.org
Information on this title: www.cambridge.org/9780521260831

First published 1990

A catalogue record for this publication is available from the British Library

Library of Congress Cataloguing in Publication data
Luzio, J. P.
Molecular medical biochemistry.
Includes index.
1. Clinical biochemistry.   2. Macromolecules –
Metabolism.   I. Thompson, R. J. (Rodney John)   II. Title.
[DNLM: 1. Biochemistry. 2. Molecular Biology.
3. Proteins.   QU 34 L979m]
RB112.5.L89   1990   612'.015   88–28541

ISBN 978-0-521-26083-1 hardback
ISBN 978-0-521-27828-7 paperback

Transferred to digital printing 2007

CONTENTS

Preface   ix
Introduction   xi

1 Proteins in medicine   1
Protein structure, including protein folding and
   domains   1
How to purify a protein   12
Analysing protein mixtures   17
Serum proteins   20
α_1-antitrypsin deficiency   23
The acute phase response   26

2 Antibodies in medicine   29
Structure and properties of antibodies   29
Antibodies as specific measuring reagents   32
Monoclonal antibodies   38

3 Tissue-specific proteins   53
Identification of tissue-specific proteins   53
The cytoskeleton   56
Microtubules   56
Microfilaments   59
Actin-binding proteins   59
Intermediate filaments   60
Intermediate filaments and disease   62
Tissue-specific enzymes in medicine   63
Isoenzymes   65
Enzymes as diagnostic reagents   71
Enzymes as therapeutic agents   73

4 The plasma membrane   79

The plasma membrane as the link between the cell and
its environment   79

The plasma membrane as an osmotic barrier surrounding
cells   79

The lipid bilayer as the basic structure of the
plasma membrane   80

The lipid components of the plasma membrane   82

Lipid asymmetry in the plasma membrane   84

The protein components of the plasma membrane   86

Protein asymmetry in the plasma membrane   90

Electron microscopy and the realization of the
universal bilayer structure of biological membranes   91

Subcellular fractionation and the preparative
isolation of membranes   95

The fluid mosaic model of the plasma membrane   99

Lateral diffusion of phospholipids   100

Lateral diffusion of proteins   100

Membrane domains and the interaction of membrane
components with cytoskeletal proteins   102

Membrane synthesis and turnover   105

Endocytosis and the recycling of membrane components   110

Cholesterol uptake: receptor-mediated endocytosis of
low-density lipoprotein   114

Transcytosis   118

Transport across the plasma membrane   121

Disease and the plasma membrane   125

5 Hormone action   128

Intercellular communication   128

Hormones as chemical messengers   128

Experimental investigation of hormone action and
hormone binding   132

Cyclic AMP as a hormone second messenger   135

The coupling of hormone-receptor binding to
activation of adenylate cyclase   137

Signal amplification and the control of metabolic
pathways   142

The control of lipolysis   145

The control of glycogen breakdown and synthesis   149

Other second messengers   154

The mechanism of insulin action 157
Long-term effects of hormones (steroid hormone
    action) 161
Hormone–hormone interactions (thyroid disease) 163

6 Secretion 166
The variety of secretory processes 166
Constitutive and triggered pathways of secretion 167
The intracellular pathways of synthesis and secretion
    of proteins 169
The signal hypothesis 170
The intracellular sorting of proteins to be secreted 172
The synthesis and intracellular packaging of insulin
    in pancreatic β cells 173
The mechanism of triggering insulin secretion 179
The origin of diabetes 185

7 The genome 188
Genes and disease 188
DNA, RNA and the genetic code 191
The structure of the human genome 196
Transformation 201
Cutting and splicing DNA 203
Techniques of DNA cloning: how to make a library 205
DNA sequencing 211
Sequence organization of the human genome 219
Transcription and transcriptional units 222
Complex transcriptional units 225
Control of gene expression 227
Splicing of primary RNA transcripts 228
Duplicated protein-coding genes and pseudogenes 231
'Defects' in eukaryotic transcription units 232
Gene probes and restriction fragment length
    polymorphisms (RFLPs) 235
RFLPs in the diagnosis of human disease 238
Hypervariable regions in the human genome 241
Genetic screening and gene therapy 242
Retroviruses and oncogenes 250

Further reading 257
Index 259

PREFACE

Biochemistry is taught to medical students at an early stage in their pre-clinical education. Often it seems less relevant to their future than other disciplines such as anatomy and physiology, and sometimes merely a course and examination hurdle to be overcome and quickly forgotten. Yet, despite the fact that it is biochemistry which so often loses teaching time to the encroachment of newer educational requirements (sociology, genetics, statistics, etc.), it is the equipment (centrifuges, scintillation counters, chromatography apparatus), and often ideas associated with biochemistry, not the accoutrements of anatomy or sociology, that fill the laboratories of departments of surgery and medicine. In accepting the opportunity to write this book, we felt we might persuade students that understanding recent developments in cellular and molecular biology will offer them as future clinicians the prospect of more rational diagnosis and therapy. In addition, we hope that a few students will be sufficiently enthused to consider research careers, applying the growing understanding of fundamental biochemical processes to a study of the common diseases of Man.

Writing even a short book requires a large amount of effort, help and advice from people other than the authors. We wholeheartedly thank our friends and colleagues, who have provided illustrations, read chapters, made useful comments and encouraged us in our endeavour. In addition, we thank Janet Eastwell, Lee Creswell and Diane Brown for typing the manuscript, and Fay Bendall and Alan Crowden of Cambridge University Press who never complained of the delays and inadequacies of our written material. We also thank those undergraduate students of the University of Cambridge whom we have supervised over the last decade (particularly at Pembroke, King's, Clare and Corpus Christi Colleges), and who have helped us inadvertently in the selection of topics for this book.

Finally, we thank especially Toni Luzio, who was extremely supportive throughout the writing.

J. P. LUZIO
R. J. THOMPSON   *Cambridge*

INTRODUCTION

This book is aimed at pre-clinical medical students studying biochemistry. In it, we have tried to give a basic account of those biochemical topics which we feel are most relevant to common clinical disorders in advanced industrial countries, even if the causes of such diseases do not yet permit complete biochemical explanation.

The health problems facing practising clinicians depend greatly on geographical location. World-wide there are 350 million cases of malaria per year (killing a million children in Africa alone), and at least 200–300 million people are infected with schistosomiasis, but neither of these diseases is prevalent in Western Europe. Moreover, the impact of infectious diseases has been greatly diminished in the past century by the development of good sanitation (and clean water supplies), vaccination programmes and the use of antibiotics. Nowadays, the common medical problems are exemplified by the histogram below which shows the main causes of general medical admissions to hospitals in England and Wales in a single year, with coronary artery disease, strokes, overdoses, diabetes and lung diseases of various sorts accounting for over half of the presenting clinical problems. It should be noted that the histogram excludes psychiatric admissions (a field where biochemistry, as opposed to pharmacology, has made little impact on diagnosis or therapy) and surgical admissions, where biochemistry is possibly least relevant.

The common diseases shown in the histogram are 'multifactorial', i.e. the end result of interaction between the products of many genes and environmental influences. The contribution of biochemistry to modern hospital practice in the treatment of such diseases is almost exclusively one of measurement, largely of substances in the serum or plasma of ill patients, sometimes on other body fluids or tissue samples and occasionally on well people. It is staggering to realize that an average-sized district hospital biochemical laboratory will carry out up to a million individual biochemical measurements per year, of which approximately 40 % will

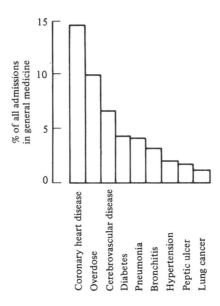

involve the use of an antibody or enzyme as a specific assay reagent. The majority of other measurements will be of simple ions or gases using physical methods. A significant proportion (10–20 %) of the work load will be quality control, i.e. the laboratory continually monitoring its own performance in the accuracy and precision of the measurements being made.

The contents of this book reflect the present use of analytical biochemical techniques to diagnose disease. Thus, in the early chapters, we describe some of the properties of proteins including antibodies and enzymes that make a particular contribution to the power of biochemical techniques in clinical diagnosis. Then we have selected topics (membrane structure and function, hormone action and secretion) where we believe current knowledge of macromolecular and cellular structure and function is likely to contribute most to understanding the underlying pathology of common multifactorial diseases. Finally, we have tried to outline the basic features of recombinant DNA techniques and the impact they are having in the diagnosis, treatment and knowledge of disease processes.

Throughout this book we assume that the reader will have access to one of the many comprehensive pre-clinical biochemistry textbooks. However, we hope we have written sufficiently clearly to preclude the need for constant reference to such a text. It is also our hope that the reader becomes aware of the excitement of recent developments in cell and molecular biology and their potential impact on the practice of medicine.

1

Proteins in medicine

Protein structure, including protein folding and domains

The human genome contains sufficient information to code for between 30 000 and 50 000 proteins. However, the majority of these (90–95 %) remain undiscovered and only a few per cent of the possible total number of human proteins are expressed at any one time in a particular cell type. Some of these proteins perform a structural role (e.g. those forming the cytoskeleton), others are enzymes catalysing and regulating metabolic pathways (e.g. the enzymes involved in glycolysis). The complement of proteins expressed by a particular cell determines its shape and functional capacity. How the diversity of cell types is directed and regulated (differentiation) remains a central problem in biology.

Proteins are large molecules varying in molecular weight from 1 to 1000 kDa and containing a linear sequence of amino acids covalently linked by peptide bonds. There is no clear dividing line between a peptide (a short chain of two or more amino acids) and a protein (a larger chain of amino acids usually with 100 amino acids or more). Chains of two or three amino acids are termed dipeptides or tripeptides, respectively, longer chains are often termed oligopeptides, and longer chains still polypeptides. Albumin in serum has 584 amino acids in a single linear chain; insulin has 51 amino acids contained in two polypeptide chains. All of these amino acids (with the exception of proline which is an imino acid) have the basic formula shown in Fig. 1.1(*a*) and differ only in the nature of the R group or side-chain. The 20 amino acids commonly found in proteins are listed in Fig. 1.2, together with the abbreviated symbols used to represent them. Amino acids occur naturally as *D*- or *L*- optical isomers, although only *L*-isomers are found in proteins. The nature of the R group determines the physical properties of the amino acid, and the polarity of the R group at pH 7.0 (which is close to physiological pH values) is particularly important. Some amino acids have negatively charged R groups at this pH (glutamic, aspartic), some have positively

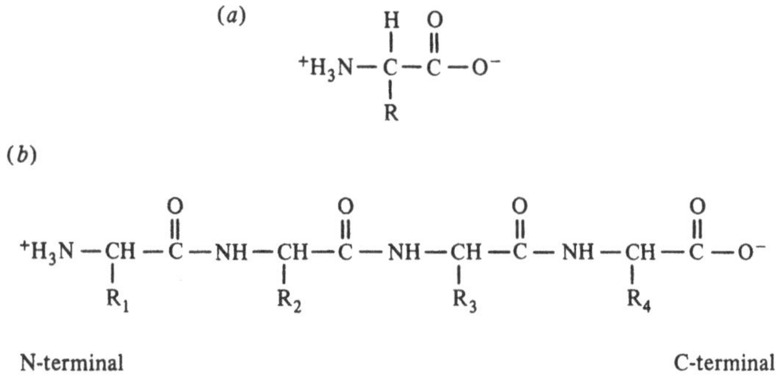

(a)

$$^+H_3N-\underset{\underset{R}{|}}{\overset{\overset{H}{|}}{C}}-\overset{\overset{O}{\|}}{C}-O^-$$

(b)

$$^+H_3N-\underset{\underset{R_1}{|}}{CH}-\overset{\overset{O}{\|}}{C}-NH-\underset{\underset{R_2}{|}}{CH}-\overset{\overset{O}{\|}}{C}-NH-\underset{\underset{R_3}{|}}{CH}-\overset{\overset{O}{\|}}{C}-NH-\underset{\underset{R_4}{|}}{CH}-\overset{\overset{O}{\|}}{C}-O^-$$

N-terminal   C-terminal

Fig. 1.1. (*a*) The basic structure of an amino acid. The R group varies according to the specific amino acid. At physiological pH the amino and carboxyl groups are charged as shown. (*b*) Amino acids are polymerized via peptide bonds between the amino group of one amino acid and the carboxyl group of the preceding amino acid. A tetrapeptide is shown. Note that polypeptide chains therefore have a free amino group at one end (the N-terminal end) and a free carboxyl group at the other end (the C-terminal end).

charged R groups (arginine, lysine, histidine), and others have uncharged R groups (glycine, see Fig. 1.2). The proportion of 'acidic' amino acids (negatively charged) (i.e. content of glutamate and aspartate) versus 'basic' amino acids (arginine, lysine, histidine) (positively charged) contained within a protein determines its overall charge at physiological pH, the total number of amino acids in the chain determines the overall molecular weight of the protein, and the order of amino acids determines the sequence or 'primary structure' of the protein. The 20 amino acids occurring in proteins can be arranged in a vast number of different sequences: for instance, a small polypeptide with each known amino acid occurring once can be arranged in 10^{18} different sequences; taking a protein of a relatively modest molecular weight of 35 kDa containing the 20 amino acids in equal numbers, and assembling one molecule of each possible sequence would produce a mass of protein greater than the weight of the earth.

Proteins exist in solution and within cells not as disordered linear sequences of amino acids but rather as folded structures with a molecular 'anatomy' characteristic for each protein. Two major 'anatomical' classes of protein are recognized, 'globular' proteins and 'fibrous' proteins. The latter, such as keratin found in hair and collagen found in tendons, are long, stringy, flexible molecules and are usually water-insoluble. Globular proteins, however, are tightly folded, water-soluble structures and it is to this class that nearly all enzymes, antibodies, serum proteins, regulatory proteins and protein hormones belong. The molecular anatomy of an individual protein is determined by the sequence of amino acids contained within its primary structure. As an example of the degree of folding of the primary structure seen in globular proteins, the 584 amino acids of serum albumin would measure 200 nm × 0.5 nm if in an extended linear array, while the circulating serum albumin molecule actually measures 13 nm × 3 nm. This folding is initially achieved by two structural motifs, the α-helix and the β-pleated sheet (both of which were first discovered as structural features of fibrous proteins). The α-helix (Fig. 1.3) forms because of the partially double-bond character of the C–N bond which cannot rotate freely, while the adjacent C–C and N–C bonds can (Fig. 1.3). It also allows hydrogen bonding between the H atom attached to the electronegative N atom of the peptide linkage and the O atom of the carboxyl group of the fourth amino acid behind it (Fig. 1.3). The 'repeat unit' of the α-helix is a single turn of 0.54 nm, representing 3.6 amino acid residues. While the α-helix represents a stable preferred conformation for a chain of amino acids linked by peptide bonds, some constraints on its formation are provided by the particular amino acids involved. For example, proline cannot take part in α-helix formation since its N atom is part of a rigid ring and cannot rotate, and no hydrogen bond is possible between proline and a partner amino acid in the α-helix since the N atom in a proline ring does not have a substituent hydrogen atom. Proline therefore always interrupts an α-helix and produces a change of direction in the polypeptide chain. Highly charged amino acids (whether negatively charged, as with glutamic or aspartic, or positively charged, as with arginine or lysine) can also prevent α-helix formation if they are repeated too frequently in the amino acid sequence because of the electrostatic repulsion by like charges. Certain other amino acids, such as leucine and threonine, also tend to prevent α-helix formation if they occur too frequently in a short stretch of sequence because of the size and shape of their R groups.

The second major arrangement of the primary structure of protein into an immediately higher-order structure is the β-pleated sheet. Here the polypeptide chain is more extended than in the α-helix and there are no intra-chain hydrogen bonds but rather inter-chain hydrogen bonds (Fig. 1.4). β-pleated sheets can form only in parts of the amino acid sequence where the R groups are relatively small. Proteins with extensive β-pleated sheet structures tend therefore to have high contents of glycine and alanine. The inter-chain hydrogen bonds between β-pleated sheets can be found between stretches of amino acid sequence running either in the

Non-polar (hydrophobic) R groups

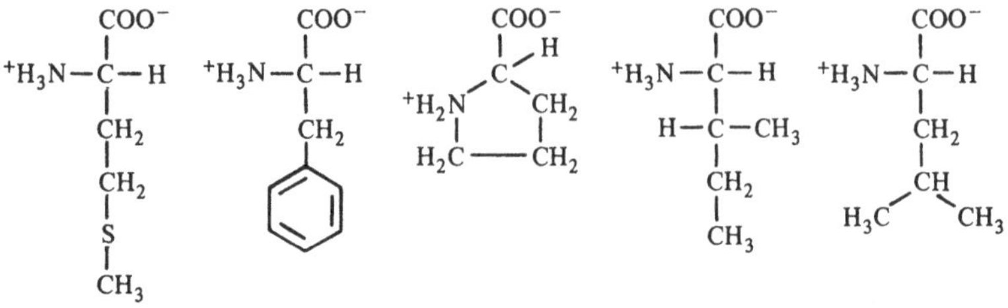

Methionine	Phenylalanine	Proline	Isoleucine	Leucine
(Met or M)	(Phe or F)	(Pro or P)	(Ile or I)	(Leu or L)

Positively charged R groups

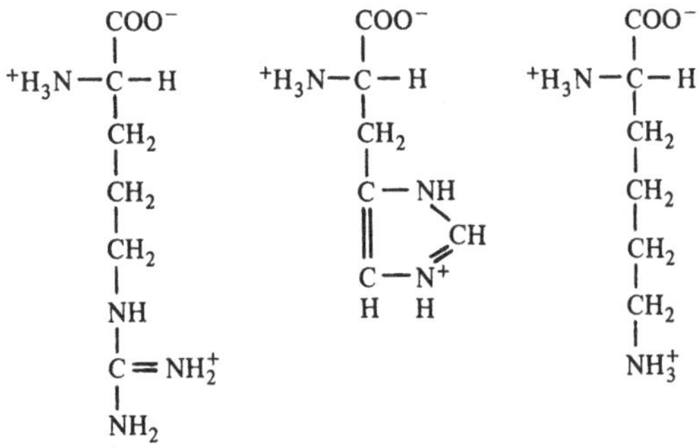

Arginine	Histidine	Lysine
(Arg or R)	(His or H)	(Lys or K)

Negatively charged R groups

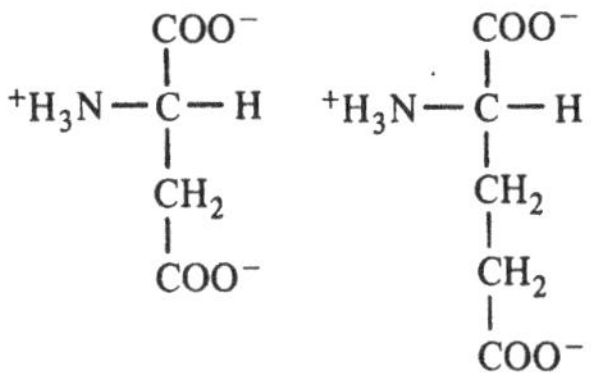

Aspartic acid	Glutamic acid
(Asp or D)	(Glu or E)

Polar but uncharged R groups

$$\begin{array}{ccccc}
\text{COO}^- & \text{COO}^- & \text{COO}^- & \text{COO}^- & \text{COO}^- \\
| & | & | & | & | \\
^+\text{H}_3\text{N}-\text{C}-\text{H} & ^+\text{H}_3\text{N}-\text{C}-\text{H} & ^+\text{H}_3\text{N}-\text{C}-\text{H} & ^+\text{H}_3\text{N}-\text{C}-\text{H} & ^+\text{H}_3\text{N}-\text{C}-\text{H} \\
| & | & | & | & | \\
\text{CH}_2 & \text{H} & \text{CH}_2\text{OH} & \text{H}-\text{C}-\text{OH} & \text{CH}_2 \\
| & & & | & \\
\text{CH}_2 & & & \text{CH}_3 & \\
\end{array}$$

Glutamine	Glycine	Serine	Threonine	Tyrosine
(Gln or Q)	(Gly or G)	(Ser or S)	(Thr or T)	(Tyr or Y)

Amino acid	Three-letter code	One-letter code
Alanine	Ala	A
Arginine	Arg	R
Asparagine	Asn	N
Aspartic acid	Asp	D
Cysteine	Cys	C
Glutamine	Gln	Q
Glutamic acid	Glu	E
Glycine	Gly	G
Histidine	His	H
Isoleucine	Ile	I
Leucine	Leu	L
Lysine	Lys	K
Methionine	Met	M
Phenylalanine	Phe	F
Proline	Pro	P
Serine	Ser	S
Threonine	Thr	T
Tryptophan	Trp	W
Tyrosine	Tyr	Y
Valine	Val	V

Fig. 1.2. The common amino acids found in proteins. The table shows the names of the amino acids with a three-letter code which is now being superseded by the one-letter code also shown. Examples of non-polar (hydrophobic), polar (uncharged), and polar (charged) groups are shown and also the structure of proline, which is an imino acid.

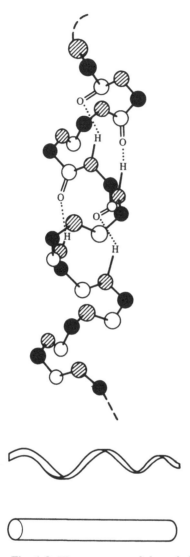

Fig. 1.3. The structure of the α-helix. Note the intra-chain hydrogen bond between the hydrogen atom attached to the nitrogen atom taking part in the peptide bond and the oxygen atom in the carboxyl group in the fourth amino acid above it. Here the R groups, which are on the outside of the α-helix, have been omitted. ● α-carbon atoms, ○-β-carbon atoms, ◉ nitrogen atoms. α-helices can be shown in diagrammatic representations of protein structures as spirals or cylinders.

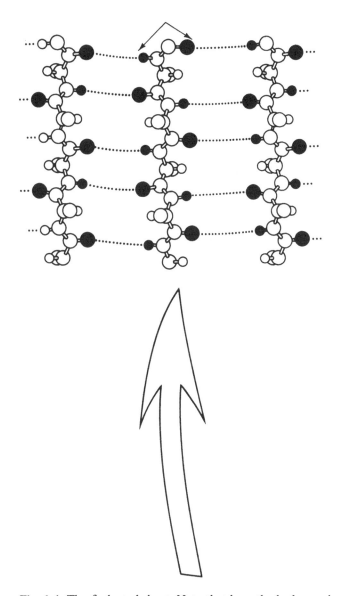

Fig. 1.4. The β-pleated sheet. Note that here the hydrogen bonds are inter-chain rather than intra-chain. β-pleated sheets can be represented in protein structures as flat arrows.

same direction (parallel β-pleated sheets) or in opposite directions (anti-parallel β-pleated sheets). The α-helices and β-pleated sheets represent the 'secondary structure' into which the primary structure is initially folded and these represent the only major extended structural elements present in globular proteins. Further folding of the protein molecule into a tightly packed globular form requires interaction between α-helices and β-pleated sheets, these structures being joined by regions of amino acid sequence not containing these organized structural motifs. This further degree of protein folding is termed 'tertiary structure' (Fig 1.5). The tertiary structure of a globular protein is maintained by four types of interaction:

1. The hydrophobic R groups of amino acids such as phenylalanine, leucine, isoleucine, methionine, valine and tryptophan seek the non-aqueous environment in the interior of the globular protein, while hydrophilic amino acids such as glutamic, aspartic, arginine, lysine and histidine are found on the outer surface of the protein in contact with the aqueous environment.
2. Electrostatic attraction occurs between positive charges on amino acids such as lysine and arginine with the negative charges on amino acids such as glutamic and aspartic.
3. Hydrogen bonding, apart from that involved in stabilizing the α-helix and β-pleated sheet structures, occurs between amino acids on adjacent loops of the polypeptide backbone.

The above three types of non-covalent forces are individually weak but collectively strong.

4. A further covalent cross-link – the disulphide bridge – occurs between intra-chain cysteine residues on adjacent loops of the polypeptide chain. It is generally believed that disulphide bridges do not occur in intracellular proteins, but rather that they are structural features of extracellular and cell membrane proteins. Albumin and other serum proteins, digestive enzymes, polypeptide hormones such as insulin, and immunoglobulins all contain disulphide bonds.

Study of the tertiary structure of several proteins (largely determined by X-ray crystallography) has shown that most proteins can be divided into five structural classes:

1. Class I proteins, or 'all α-proteins', in which only α-helices are present;
2. Class II proteins, in which only β-structures are present, usually as anti-parallel strands;
3. Class III proteins, in which both α and β structures are present but in different and separate areas of the protein molecule;
4. Class IV proteins (or α/β proteins), in which α and β structures

alternate in the primary structure, usually giving a central region with parallel β-sheets with exterior flanking α-helices; and

5. Class V proteins, which are usually small, show little secondary structure, and are often rich in disulphide bridges or have a large associated cofactor.

Numerous proteins are now known to have 'domains' within their tertiary structure. These are most commonly 100 to 200 amino acids in length and are folded into locally independent complete structures looking like small, intact protein molecules in their own right (Fig. 1.6). Domains within the whole protein molecule are linked by stretches of polypeptide chain with a less ordered structure. It is becoming increasingly clear that domains within proteins can move relative to each other and that this can be related to the biological role of the protein. For example, enzymes with domain structure nearly always have their active sites at the interface between domains, and the binding of substrate

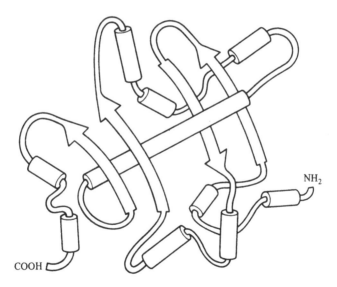

Fig. 1.5. The anatomy of a protein. α-helices (represented here as cylinders) and β-pleated sheets (broad arrows) are connected by regions of randomly structured polypeptide chain, the whole protein molecule being held in a specific three-dimensional structure. The interior of the structure is hydrophobic; charged R groups mainly face the hydrophilic outer surface of the protein.

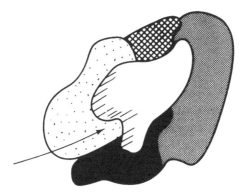

Fig. 1.6. Protein domains. Portions of the protein chain form local domains appearing as circumscribed local areas of tertiary structure which can move relative to other domains. Active structures of enzymes (e.g. as arrowed here) are frequently at domain junctions.

molecules can induce the domains to 'close', as in the binding of glucose by hexokinase. The ability of domains to move is undoubtedly important in enzyme action, allosteric control, and the assembly of proteins into larger structures, e.g. the construction of virus coats from repeating units of identical proteins.

The folding of a newly synthesized polypeptide chain of average length (300–500 amino acids) into a compact globular structure within the cell occurs rapidly (within seconds or minutes) and it has long been realized that randomly 'trying-out' all possible conformations until the right one is found cannot possibly explain the mechanism of protein folding. The widespread presence of domains in globular proteins suggests that they represent smaller folding units, making possible the rapid attainment of the correct tertiary structure. Isolated domains (i.e. portions of the amino acid backbone normally constituting a domain which have been excised by using proteolytic enzymes) can be shown experimentally to fold up spontaneously. A current view of the process of protein folding is that 'nucleation' sites form in different areas of the polypeptide chain. These may differ with different types of protein, but could be the local formation of α-helices or of β-strands, or possibly the clustering of a hydrophobic group of amino acids. These areas of organization associate by non-covalent forces, throwing out loops or stretches of less ordered structure to allow, for instance, α-helices and β-strands to come together. This is then readjusted to form a stable overall structure and disulphide bridges (if they are to be present) are added. The final stage is the association of new, relatively rigid domains into an overall compact

globular protein. Large proteins (of 50 kDa and above) are often composed of two or more polypeptide chains folded as above but associated
non-covalently into a higher order or 'quaternary structure'. These
proteins – examples are haemoglobin and most of the enzymes catalysing
reactions in the glycolytic pathway – therefore have a 'subunit' structure.
Two subunits (a dimeric protein) or four subunits (a tetrameric protein)
are common; proteins made up of three subunits (trimers) are unusual.
Allosterically regulated enzymes nearly always have a subunit structure.
New recombinant DNA techniques are making it possible to substitute
different amino acids in key structural areas of proteins ('site-directed
mutagenesis'), and this should throw light on the ways in which primary
structure determines tertiary and quaternary structure.

Many proteins are further modified after the assembly of the primary
amino acid sequence by the ribosome. A common 'post-translational'
modification is the addition of carbohydrate side-chains to serine or
threonine groups (O-linked sugar chains) or to asparagine (N-linked
sugar chains). These proteins are now 'glycoproteins' and such modification is very common in secreted proteins and in membrane proteins which
penetrate to the outside of the cell. The carbohydrate side-chains alter
the solubility of secreted proteins and may act as signals determining the
length of time the protein survives in the circulation before being
degraded. Cytoplasmic proteins are not glycosylated. Other post-translational modifications include phosphorylation of serine, threonine and
tyrosine residues (an important theme in the control of enzyme activity by
hormones – see Chapter 5), the attachment of fatty acids to cysteine
residues in membrane proteins, and the activation of proteins by proteolytic cleavage, common in the clotting factor cascade and in the
production of digestive enzymes in the gut. Many proteins have an acetyl
group covalently added to the N-terminal amino group, but the biological
function of this N-terminal modification is unclear.

The amino acid sequences of proteins can be used to gain insight into
their evolutionary history. The extent to which the same amino acids
occur in the same places in the sequence of two different proteins
provides a clue as to whether they are derived from a common ancestral
protein. Some proteins, e.g. haemoglobin and myoglobin, show high
degrees of homology and are clearly derived from the same evolutionary
precursor. The commonest deviation between two functionally similar
proteins from closely related organisms is the substitution of one amino
acid for another with similar physical properties so that the tertiary
structure of the protein is conserved. As species diverge, more and more
substitutions may occur until it is impossible to say whether or not the

proteins are derived from a common ancestor. The random arrangement of 20 amino acids into two different polypeptide chains should lead to the same amino acid being in the same position in about 5 % of cases. In practice, and mainly because of insertions or deletions of amino acids into protein sequences, thus altering the 'register', it is not possible to decide whether a degree of homology of less than 15 % is due to chance rather than a common evolutionary background.

The existence of introns (see p. 228) in eukaryotic genes and domains in large globular proteins has suggested a mechanism for speeding up the evolution of proteins. In several instances, the boundaries of domain structures in proteins have been shown to correspond to intron/exon boundaries in the corresponding structural genes. Recombination events inside the introns of two unrelated structural genes could transfer whole domains between proteins, resulting in the production of new proteins with biological properties derived from the domains of both 'parent' proteins. There is evidence for such 'exon shuffling' in the serine protease family (trypsin, chymotrypsin, elastase and several other proteases). However, in other proteins, e.g. haemoglobin, introns appear in the middle of the coding sequence for what is regarded as a domain in the completed protein. Furthermore, in some proteins, a single complete domain is coded for by more than one exon. The role (if any) of introns in speeding protein evolution will hopefully become clearer as more structural genes are sequenced.

How to purify a protein
A prerequisite for the sequencing of a protein and the investigation of its tertiary or quaternary structure by physical techniques such as X-ray crystallography is that the protein can be obtained in reasonable quantities (milligrams) in a pure homogeneous form, i.e. free from contamination and in an undegraded state. Often the protein of interest is an enzyme and studies of its catalytic activity, specificity and mechanism may require it to be pure. A further reason for purifying proteins is for use in raising antibodies. These make it possible for the protein to be measured specifically in complex mixtures of other proteins and also enable the protein to be localized to specific cell types within a tissue by immunohistochemical techniques.

This section summarizes the range of modern techniques used to purify one protein from the several thousand other proteins present in a human tissue (Fig. 1.7). It is necessary to have an 'assay' for the protein, i.e. to be able to detect its presence in different fractions produced by the purification procedure. This can be done by measuring its catalytic activity (if the protein is an enzyme), by following its binding activity (if the protein is a receptor of a drug or is a metal-binding protein), by detecting its presence immunologically if a specific antibody is available, or by knowing its position of migration on a one- or two-dimensional analytical gel (see below). The following basic method is for the purification of a 'typical' cytoplasmic protein soluble in aqueous solvents. The purification of hydrophobic protein components of biological membranes usually requires more difficult techniques, often employing detergents and sometimes organic solvents (see Chapter 4).

Choose a tissue with a high content of the protein of interest. Homogenize (break up) the tissue in a suitable aqueous buffer containing any known stabilizing agent, e.g. a reducing agent to keep sulphydryl groups reduced or a metal ion if this is known to preserve a particular enzyme activity. Some chemical substances, e.g. phenylmethylsulphonylfluoride (PMSF), are sometimes included to inhibit proteases released from inside cells.

Centrifuge away the cell debris, take the clear supernatant and process it by one (or all) of the following purification procedures, which represent the techniques commonly used to isolate proteins from complex mixtures:

1. ammonium sulphate precipitation;
2. gel-filtration or 'molecular sieving' (Fig. 1.7(a));
3. ion-exchange chromatography (Fig. 1.7(b));
4. isoelectric focussing;
5. affinity chromatography (Fig. 1.7(c)).

Each individual purification scheme is tailored to fit the protein of interest. Some or all of the above procedures may be used sequentially in whatever order produces the most rapid and convenient purification. The production of large quantities (i.e. milligrams or grams) of pure proteins for further analysis can be labour-intensive and time-consuming, often taking days or weeks to bring to completion.

Ammonium sulphate precipitation

Different proteins precipitate out of solution at different ammonium sulphate concentrations, a procedure known as 'salting out'. The supernatant solution is made, e.g. 20 % saturated in ammonium sulphate (usually at 0 °C). After stirring, any precipitate is removed by centrifugation, the supernatant is increased to 40 % saturated and the procedure is repeated, and so on. In this way, precipitates are obtained from ammonium sulphate 'cuts' of 0–20, 20–40, 40–60, 60–80, 80–100 %

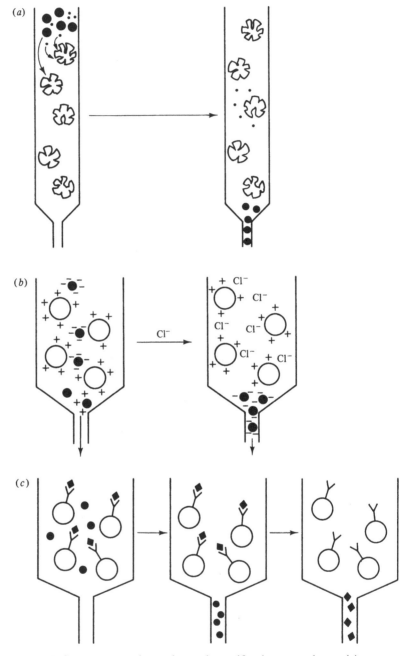

Fig. 1.7. Some commonly used protein purification procedures. (*a*) Gel-filtration or 'molecular sieving'. Dextran beads with holes in them are packed into a long column. When a mixture of proteins is passed down the column, large proteins cannot enter the beads and

saturation, and these can be analysed for the presence of the protein under investigation. Very few proteins are soluble in 100 % ammonium sulphate.

Gel-filtration or 'molecular sieving'

A column is set up containing a packed bed of 'beads'. These are often of dextran but can be of other material, e.g. polyacrylamide or even glass. Small holes penetrate the surface of each individual bead. As the mixture of protein molecules passes down the column, small proteins are able to enter these holes; larger proteins cannot and they bypass the internal volume of the beads. Larger proteins therefore emerge from the base of the column before smaller proteins, and hence proteins are separated on the basis of molecular weight (and to a less extent on the overall shape of the protein molecule). Gel-filtration columns are typically long and thin. By choosing different types of bead, it is possible to select the molecular weight range over which the column separates proteins most effectively, e.g. proteins of molecular weight less than 50 000 kDa, those of 50 000–100 000, 100 000–200 000, etc. By 'calibrating' the column with proteins of known molecular weights, it is possible to estimate reasonably accurately the molecular weight of an unknown protein by passing it down the column under the same elution conditions.

Fig. 1.7. (*cont.*)

so emerge from the column ahead of smaller proteins, which pass in and out of the beads and are therefore delayed. (*b*) Ion-exchange chromatography. A column is packed with cellulose, chemically modified to contain covalently linked positive (or negative) charges. Protein mixtures are poured into the column and 'stick' because of ionic attraction. A steadily increasing concentration (i.e. a gradient) of a competing ion (e.g. an NaCl solution) is then applied to the column. This competes with the charges on the column and the attached proteins are detached and eluted. Weakly charged proteins emerge before strongly charged proteins. (*c*) Affinity chromatography. A solid support (e.g. cellulose) is prepared with a specific affinity molecule (a ligand) attached. This ligand can be a particular substitute for an enzyme, an antibody directed against a particular protein (see p. 43), or a hormone or toxin reacting with a specific receptor. A mixture of proteins is passed into the column, the specific protein required binds to the affinity ligand and all other non-reacting contaminating proteins pass through. The protein can then be removed by non-specific means (e.g. pH changes or high-salt conditions) which disrupt the binding of the ligand to the required protein, or by washing with a solution of free ligand.

Ion-exchange chromatography

This process depends on the different individual total electrostatic charges present on individual proteins in a mixture. The net charge on a protein at a particular pH is determined by its content of 'acidic' and 'basic' amino acids, the charges on the amino and carboxyl terminal amino acids of the polypeptide chain usually being of minor importance. A protein rich in aspartate and glutamate residues is said to be 'acidic' and will carry a high net negative charge at alkaline pH; a protein with a high content of arginine and lysine residues is said to be 'basic' and will carry a relatively high net positive charge below pH 7.0. Between pH 8.0 and pH 9.0, the majority of human proteins carry a net negative charge. If a mixture of such proteins buffered to this pH range is passed through a resin column containing positive charges, most proteins will stick to the column by electrostatic attraction, with the more minor proportion of basic proteins passing straight through. If the column is now washed with a gradually increasing concentration of a solution containing a positive ion (e.g. an increasingly concentrated NaCl solution), the electrostatic attraction is neutralized and the protein passes into free solution and elutes from the column. The concentration of positively charged competing ion at which this occurs varies from protein to protein in the original mixture (being higher for the more acidic proteins) and individual proteins therefore elute from the column at different times. Ion-exchange columns therefore separate by differences in net charge at a particular pH and are characteristically short and fat. Mixtures of more basic proteins can be separated on negatively charged columns at pH values below 7.0. Excesses of pH, e.g. pH 3.0 or 10, often lead to denaturation of proteins.

Isoelectric focussing

Individual proteins show characteristic individual isoelectric points. This is the pH at which the protein shows no net negative charge and is determined by the proportion of acidic and basic amino acids in the protein and by the individual pKs of each residue. A pH gradient (e.g. from pH 3.0 to 9.0) is set up and the protein mixture to be separated is introduced. A voltage is set up across the pH gradient, causing each individual protein to migrate to the pH value corresponding to its isoelectric point. At this pH the protein is uncharged and therefore does not migrate further. In this way each protein is 'focussed' into a narrow pH range and proteins differing by a single charge can be easily separated.

Affinity chromatography

Many proteins, e.g. enzymes and receptors, bind smaller molecular weight 'ligands', e.g. substrates, hormones and neurotransmitters. This property can be used to purify a protein directly from a mixture. A column is set up with a solid support (e.g. finely divided cellulose or agarose) to which the appropriate ligand is covalently attached. The mixture of proteins is then passed down the column. Only the protein able to bind the specific ligand 'sticks' to the column; other proteins pass through. The desired protein can be removed from the column by washing it with a high concentration of the ligand free in solution. Affinity chromatography can also be performed using antibodies covalently attached to solid supports. These specifically recognize and bind the desired protein in the mixture being applied to the column (see p. 43).

Analysing protein mixtures

The routine analysis of mixtures of proteins to determine the number and molecular weights of different polypeptide chains present is now almost exclusively carried out using polyacrylamide gel electrophoresis (PAGE). Polyacrylamide is a polymer which can be poured as a liquid. After pouring, and with the appropriate added catalysts, it sets to a clear, flexible, solid support in which electrophoresis of proteins can be carried out. Polyacrylamide gels are either cast in glass tubes (tube gels) or, more commonly, cast between two glass plates separated by spacers and sealed from the edges (slab gels). Protein mixtures can be analysed under either 'native' (non-denaturing) or denaturing conditions. Non-denaturing PAGE systems are usually run at pH values between 8.0 and 9.0. Most proteins carry a negative charge at this pH and therefore migrate towards the positive electrode, with more acidic proteins migrating most rapidly. Under these PAGE systems, the tertiary and quaternary structures of the proteins being analysed are not destroyed, e.g. proteins composed of subunits remain intact. With denaturing PAGE, however, proteins are dissociated into their individual polypeptide chains by complexing with the anionic detergent sodium dodecyl sulphate (SDS). The SDS unfolds the polypeptide chains and associates with them in such a way that the hydrocarbon chains of SDS form a tight hydrophobic interaction with the polypeptide chain and the negatively charged sulphate groups of the SDS are exposed to the aqueous medium. The unfolded polypeptide chains contain a constant ratio of 1.4 g SDS per g protein, thus having a uniform negative charge per unit length. Different polypeptides then differ only in mass and can be separated according to molecular size in a cross-linked polyacrylamide gel when an electric field is applied. (All the proteins migrate to the positive electrode.) The technique can be used to analyse any protein mixture or to estimate the apparent molecular weight of a specific protein by comparison with known standards, although difficulties can be encountered with glycoproteins containing more than 10 % carbohydrate which tend to run anomalously. Reducing agents such as mercaptoethanol or dithiothreitol are used in conjunction with the SDS when it is necessary to separate polypeptide chains joined by disulphide bonds. By far the commonest method of estimating the molecular weights of proteins is now SDS–PAGE electrophoresis in parallel with standard proteins whose molecular weight is already known. It is also a technique particularly useful for analysing membrane proteins (see Chapter 4, Fig. 4.10).

Although SDS treatment is a good means of unfolding soluble protein and solubilizing membrane proteins, it usually results in the inactivation of their biological activities. In analytical systems, the SDS can be electroeluted away from the protein with some success, and recovery of antigenic and even enzymatic activity has been reported. A good example is the technique of immunoblotting (or Western blotting), where proteins separated on an SDS–PAGE gel are transferred in an electric field on to nitrocellulose to which they bind tightly. Much of the SDS in the SDS–PAGE gel is removed during this process. The nitrocellulose is then incubated in a buffered solution at physiological pH containing protein which blocks remaining binding sites on the nitrocellulose and allows refolding of the denatured polypeptide chains. The blotted proteins can then be challenged with antibody or enzyme substrate to define specific antigenic or enzymic proteins within a mixture (Fig. 1.8).

The resolving power of 'one-dimensional' PAGE systems under either non-denaturing conditions (proteins of the same charge migrate in similar positions) or denaturing conditions with SDS (proteins with the same molecular weight migrate in the same position) is limited. An enormous increase in the power of PAGE systems to resolve protein mixtures came with the introduction of two-dimensional systems. Here, protein mixtures are first subjected to isoelectric focussing (which separates proteins according to isoelectric point) in a thin cylinder of polyacrylamide. This cylinder is then placed on top of a slab gel containing SDS (which separates according to molecular weight) and electrophoresed in a second dimension. Since the isoelectric point and the molecular weight of a protein are unrelated parameters, the resolving power of the two-dimensional system is maximized. The gels can be stained with either Coomassie blue to show between 200 and 300 spots, or with more sensitive silver stains (based on photographic techniques) to show several

(*a*)　　　　　　　　　　　　　　　　(*b*)

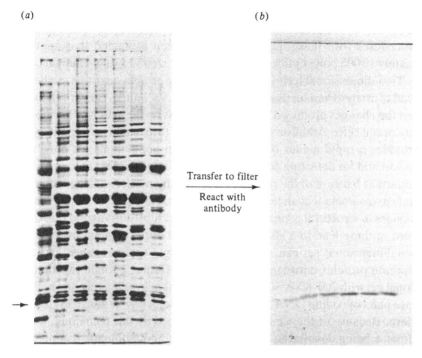

Transfer to filter
→
React with
antibody

Fig. 1.8. SDS–PAGE and Western blotting. (*a*) is a photograph of
an SDS 'slab' gel. Seven protein mixtures have been applied to the
top of the polyacrylamide slab in seven separate lanes and
electrophoresed towards the positive electrode. The gel has then
been stained to show individual protein bands. In this system, small
proteins (~10 kDa) run fastest, large proteins (~100 kDa) slowest.
The proteins have then been electrically transferred to a
nitrocellulose sheet shown in (*b*). The sheet has been treated with
an antibody directed against the arrowed protein shown in lane 1.
The antibody binds to the transferred protein and the binding is
detected by linking to a peroxidase reaction as with
immunohistochemistry (see Chapter 3). The antibody binds to a
similar protein in all lanes, showing that the same protein is present
in all the initial protein extracts which came from seven different
animals. 'Western blotting' of proteins is so-called to distinguish it
from Southern blotting (devised by E. Southern, in which DNA is
transferred and probed with DNA) and Northern blotting (in which
RNA is transferred and probed with DNA or RNA).

thousand spots, or the proteins can be made radioactive before analysis
and their presence detected by autoradiography. The standard dimensions of a two-dimensional slab gel are 15 cm × 15 cm. By using 'giant'
gels (approximately 40 cm × 40 cm), greater loadings of protein mixtures

can be applied and greater resolution achieved. Up to 3000 proteins can be seen by conventional staining methods, and by using autoradiographic procedures with these 'giant' gels it has been estimated that approximately 10 000 polypeptides are being synthesized in a single human cell.

Two-dimensional techniques for analysing protein mixtures are being used to analyse human tissue proteins (Fig. 1.9), human cancer cell lines, and the changes produced in protein synthesis by the action of hormones on normal cells. Additionally, this technique (especially with 'giant' gels) provides a rapid means of surveying human populations for polymorphisms and for detecting point mutations at a large number of loci. This is important because of the possible effects of chemicals, ionizing radiation and environmental changes. It is estimated that about 75 % of single base changes in structural genes lead to amino acid substitutions and that one-third of these lead to a change in the charge of the protein involved. A two-dimensional gel can easily detect a single charge change and can separate proteins between 100 and 10 kDa. Hence, a single two-dimensional gel with 700–800 visible spots represents approximately one million base pairs of coding DNA and is capable of detecting a change in 25 % of these. Because of the wealth of data produced by this technique, much effort is being devoted to computer analysis and automatic scanning of two-dimensional gels of human proteins.

Serum proteins
Proteins circulating in human blood are readily accessible and can be analysed directly to produce diagnostic information on disease states in patients. These analyses are carried out either on plasma (i.e. blood to which an anticoagulant has been added and then the blood cells centrifuged off) or on serum (i.e. blood which has been allowed to clot and the clear fluid above the blood clot used, hence serum is plasma minus fibrinogen). There are several hundred proteins in serum, the functions of most of which are unknown. The terms 'albumin' and 'globulin' date from the middle of the last century, 'globulin' referring to the small white precipitated particles obtained by salt treatment of serum. From the 1930s until quite recently, one-dimensional, non-denaturing electrophoretic methods have been used to analyse serum proteins; this is now most commonly done on agarose (a porous material derived from seaweed). Many textbooks refer to 'α-globulins', 'β-globulins' and 'γ-globulins', meaning those proteins which migrate in the α, β and γ regions of the electrophoretic strip. These migration positions reflect merely the net charge on the serum protein, and these terms are dying out as specific functions are assigned to more and more individual proteins. Analysis of

3.0   pH value   9.0

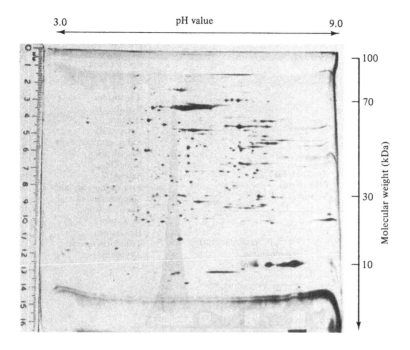

Fig. 1.9. Two-dimensional, high-resolution polyacrylamide gel electrophoresis. The soluble (cytoplasmic) proteins in human liver cells have been separated into several hundred individual polypeptide chains. Separation is by isoelectric focussing in a pH gradient in the first dimension with the acidic end of the gradient (pH 3.0) to the left as shown. The second dimension is an SDS system. (Approximate molecular weights are shown on the right.) The large streak in the upper central gel is albumin which is made in large amounts by the liver. The gel has been stained with dye (Coomassie blue) and several hundred protein spots are visible. The minimum amount of protein needed to be visible in this system is about 50 ng. The use of more sensitive detection systems, e.g. silver staining procedures, or radioactive labelling of proteins before separation, would show several thousand proteins on this gel.

serum proteins is routinely carried out by electrophoresing $2\,\mu l$ of serum. The pattern obtained is shown in Fig. 1.10. Of the several hundred proteins in serum, only about 13 are present in serum in high enough concentrations to show as discrete bands on the strip. (This is partly due to the disproportionately high concentration of albumin at $60\,g\,l^{-1}$. Putting more serum on to the strip to show up more minor serum proteins would overload the system with albumin and hence lose resolution.) Immunoglobulin molecules, IgA, IgG and IgM (see Chapter 2), do not

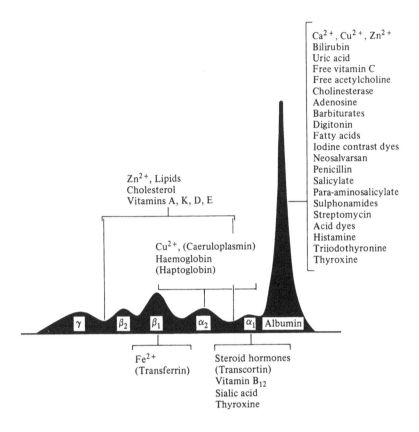

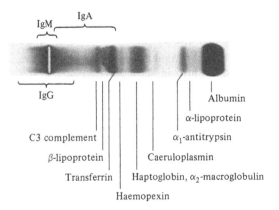

Fig. 1.10. Analysis of human serum proteins. The appearance of an electrophoretic 'strip' of $2\,\mu l$ of human serum is shown in the bottom part of the figure with visible human proteins named. The upper part of the figure illustrates the wide range of biological and non-biological substances transported in serum.

form discrete bands because they represent widely heterogeneous populations of molecules with widely varying electrophoretic mobility.

The two major functions of serum proteins are the transport of substances within the circulation, and the defence of the body against damage from within or without. The major transport proteins and the substances transported are shown in Fig. 1.10. Not only do serum proteins carry naturally occurring substances such as triglycerides (see Chapter 4), fatty acids, essential metal ions, steroid hormones, etc., from organ to organ via the circulation, but externally acquired substances such as drugs, dyes and vitamins are also tightly bound and transported by serum proteins. The three major defence functions of serum proteins are anti-protease activity, blood clotting and immune defence. An effective defence system against proteolytic enzyme activity appearing in the circulation has been evolved, mainly involving α_1-antitrypsin and α_2-macroglobulin. Several serum proteins are involved in the clotting mechanism. Immune system proteins comprise the complement proteins and the five classes of immunoglobulin molecule (see Chapter 2).

The major reasons for analysing serum proteins are threefold. First, nearly all of the serum proteins shown in Fig. 1.10 show inherited variations or polymorphisms, sometimes resulting in deficiency or actual absence of the protein from the circulation. The great majority of inherited variations in serum proteins produce no clinical symptoms and are mainly of use in studies of the genetic make-up of different populations and races. Deficiencies of serum proteins have variable clinical impact, depending on the particular protein involved. Below, we shall consider a well-defined example of such deficiency involving the protease inhibitor α_1-antitrypsin. The second main reason for analysing serum proteins in ill patients is to establish the presence of an 'acute phase reaction' which is a good indication of tissue damage and which involves the study of a group of serum proteins known as 'acute phase' proteins (also considered in more detail below). The third major reason is to detect the presence of 'paraproteins' in the circulation. These represent the abnormal production of 'monoclonal' immunoglobulins by the immune system and will be dealt with in Chapter 2.

α_1-antitrypsin deficiency

In 1964, Laurell and Erikson in Sweden were studying the levels of the anti-protease inhibitor protein α_1-antitrypsin in the serum of patients with lung disease and found a striking absence of α_1-antitrypsin in the serum of some patients with pulmonary emphysema (Fig. 1.11). This is a disease in which the elastic tissue between the normal air sacs of the lung break

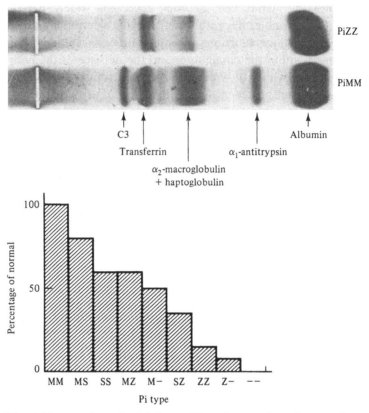

Fig. 1.11. α_1-antitrypsin deficiency. Note the complete absence of a visible α_1-antitrypsin in the PiZZ patient. Other Pi variants shown in the lower part of the figure also have lowered α_1-antitrypsin levels, although it is not clear how clinically relevant these less severe variants are. Reproduced with permission from *Trends in Biochemical Sciences*, Elsevier , Cambridge.

down and coalesce into larger spaces, thus greatly reducing the surface area available for gaseous exchange and sometimes leading to crippling breathlessness. This dramatic demonstration of a link between the absence of a serum protein and a well-defined lung disease was a landmark in molecular pathology and stimulated much interest in human variants of α_1-antitrypsin. It is known that there are at least 30 inherited variants of α_1-antitrypsin in humans. These are classified by the Pi (protease inhibitor) system and range from the PiMM phenotype with normal (100 %) serum levels of α_1-antitrypsin to the severe α_1-antitrypsin deficiency originally discovered by Laurell and Erickson (the PiZZ variant in which levels of 5 % or less of α_1-antitrypsin are found in the circulation). Different variants of human α_1-antitrypsin between these two extremes with intermediate serum levels of α_1-antitrypsin are also known (e.g. PiSS, PiSZ). Most human α_1-antitrypsin variants with low serum levels of the protein do not have the α_1-antitrypsin deleted but rather synthesize an abnormal form of the protein, e.g. in the Z form of α_1-antitrypsin a glutamate residue at position 342 in the amino acid sequence is replaced by a lysine residue. This results in the hepatocyte being unable to process and secrete the protein into the circulation, and the consequent accumulation of large amounts of α_1-antitrypsin in the liver.

The lung disease (emphysema) associated with PiZZ α_1-antitrypsin deficiency develops because the lungs are a site of destruction of white cells from the circulation. These contain elastase, an enzyme which can digest the elastin in the walls of the alveolar spaces. In the normal person, the presence of α_1-antitrypsin in the circulation protects the elastic air spaces by binding and inactivating the released elastase.

α_1-antitrypsin is a highly ordered molecule with 40 % of its structure in β-pleated sheets and 30 % in an α-helical configuration. The 'reactive centre' of the molecule is an exposed loop of 16 amino acids which is believed to hold the whole molecule in a stressed metastable state (Fig. 1.12). The sequence around methionine 358 is Pro–Met–Ser–Ile, which fits into the active site of neutrophil elastase and the Met–Ser bond is cleaved. This cleavage results in the formation of a tight 1:1 complex between the elastase and the cleaved α_1-antitrypsin molecule, which is removed from the circulation and broken down. Thus, the exposed reactive centre of α_1-antitrypsin acts as a 'bait' for elastase and is known as a molecular 'mousetrap', a phenomenon also seen in other anti-protease inhibitors. For example, mutation of methionine 358 to arginine

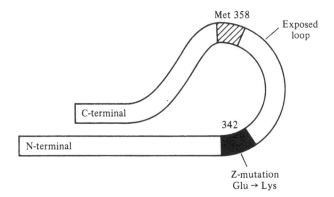

Fig. 1.12. The 'molecular mousetrap' of α_1-antitrypsin.

results in a protein specifically inactivating thrombin rather than elastase, a condition leading to a bleeding tendency in the patient carrying the mutation.

In the α_1-antitrypsin deficient PiZZ individual, the unprotected air spaces are broken down. The accumulation of the abnormal $Z\alpha_1$-antitrypsin protein in the liver leads to a second feature of this disease, namely cirrhosis. This is a fibrous 'scarring' of the liver (more often seen as a consequence of excess alcohol intake) which eventually destroys the normal population of liver hepatocytes and leads to liver 'failure'. Characteristically, the cirrhosis in α_1-antitrypsin deficient patients has its onset in early life and is known as 'juvenile' cirrhosis. One form of α_1-antitrypsin deficiency – the Pi 'null' variant – is known in which the gene is apparently absent, rather than present and coding for an abnormal protein as with other forms of α_1-antitrypsin deficiency. Since no abnormal protein accumulates in the liver, these patients do not develop juvenile cirrhosis but still develop emphysema in later adult life as the lung air spaces remain unprotected.

α_1-antitrypsin deficiency is one of the commonest inborn errors in Western populations, with 5 % of Scandinavians, 4 % of Britons and 3 % of Americans heterozygous with the MZ phenotype. Emphysema, is a very common disease, and only a few per cent of cases can be shown to be due to PiZZ α_1-antitrypsin deficiency. However, a high proportion (at least half) of the patients with PiZZ α_1-antitrypsin deficiency will develop emphysema. PiZZ α_1-antitrypsin deficiency is largely confined to European races and is not found in Negroes (hence the disease has been called the 'white man's sickle-cell anaemia'). The Z mutation is thought to have arisen in a single North European individual 6000 years ago, but the reason for the maintenance of this gene in European populations, presumably confirming a selective evolutionary advantage, is not clear, although many speculative explanations have been proposed.

The acute phase response

The acute phase response is an orchestrated pattern of changes seen in the serum proteins of ill patients soon after (hence 'acute' phase) tissue damage produced by a wide possible variety of causes. The existence of this response to tissue damage was first recognized in the 1930s in the laboratory of Avery, where later the significance of DNA as the carrier of genetic information was realized. Patients with pneumococcal pneumonia were found to have a protein in their serum which would bind to a polysaccharide from the pneumococcal cell wall called 'C-polysaccharide', which is why the protein was called C-reactive protein or CRP (see Fig. 1.13). At first, this protein was thought to represent some primitive 'antibody' response but, unlike normal antibodies, CRP was present in serum within hours of the onset of the illness and then disappeared as the patient's temperature fell and convalescence began. Normal antibodies behaved in the opposite fashion, i.e. they appeared late in the disease and increased in serum level (titre) as convalescence began. It was soon found that the appearance of CRP in the serum of ill patients was not confined to patients with pneumococcal pneumonia but that the protein also appeared rapidly in the serum of patients with a wide variety of other conditions causing tissue damage, e.g. other types of infection, myocardial infarctions (heart attacks), following surgical operations, etc. CRP is now known to be the archetypal 'acute phase reactant' protein; other acute phase reactant proteins are shown in Fig. 1.13. In the acute phase, the production of some acute phase proteins, e.g. CRP itself and α_1-antitrypsin, is increased rapidly (within 6 h) and dramatically (up to 1000-fold) by the liver. These are 'positive' acute phase reactants. Other proteins (e.g. albumin) are produced at lower levels than normal and are therefore 'negative' acute phase reactants. In some cases, e.g. α_1-antitrypsin (an anti-protease inhibitor) and haptoglobin (which binds released haemoglobin from lysed red cells), the increased synthesis of 'positive' acute phase proteins makes biological sense. In other cases, e.g. the increased synthesis of α_1-acid glycoprotein (orosomucoid) and CRP itself, the biological role of the protein is not understood, although it has recently been suggested that CRP may function in binding to chromatin fragments released from damaged cells. CRP has been sequenced and is a pentamer of five identical subunits, showing no homology with the primary amino acid sequence of

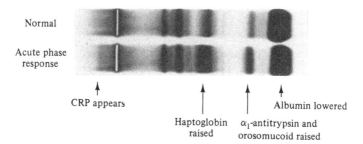

Fig. 1.13. A normal serum electrophoretic strip and one from a patient with an acute phase reaction following a myocardial infarction. Reproduced with permission from *Trends in Biochemical Sciences*, Elsevier, Cambridge.

immunoglobulins. The importance of the acute phase response clinically lies in the measurement of acute phase protein levels (especially CRP, α_1-antitrypsin, haptoglobin and orosomucoid) in the serum of acutely ill patients. The stimulus to the liver protein synthetic machinery to increase (or decrease) the synthesis of acute phase proteins is thought to be mediated by 'kinins', released from damaged tissues, which pass to the hepatocyte via the circulation. The healing of tissues 'switches off' the acute phase response and the return of serum acute phase protein levels to normal provides the clinician with good evidence of the patient's recovery.

2

Antibodies in medicine

Structure and properties of antibodies

The immunoglobulin molecule consists of two heavy and two light chains linked by disulphide bridges (Fig. 2.1). Each heavy and light chain has a constant and a variable region, and the variable regions of both heavy and light chains take part in forming the specific binding sites of the antibody, of which there are two per immunoglobulin (antibody) molecule. Heavy chains in the immunoglobulin molecule define the class of the antibody, i.e. α chains in IgA, δ chains in IgD, ε chains in IgE, γ chains in IgG, μ chains in IgM. IgM is usually present in serum as a pentamer (Fig. 2.1(b)) joined by special bridge peptides (J chains). Similarly, IgA can exist as a dimer (known as polymeric IgA) in human serum (see Chapter 4, Fig. 4.30).

Within each class of antibodies circulating in the body, there are many different individually distinct immunoglobulin molecules directed against a single antigenic determinant or 'epitope'. Each individual immunoglobulin molecule is produced by a single clone of plasma cells which has been programmed by the immune system to respond to a particular antigenic determinant. During the primary and secondary immune responses (Fig. 2.2), many different clones of plasma cells respond to different epitopes (antigenic determinants) on the injected antigen, and hence the normal immune response is 'polyclonal'. With protein antigens (which form the majority of antigens of biological interest), epitopes that give rise to an immune response can be either continuous (i.e. composed of a short stretch of, say, five or six amino acids in the primary sequence) or discontinuous (i.e. formed from amino acids separate in the primary sequence but brought into close proximity by features of secondary or tertiary structure). Current evidence suggests that antigenic 'hot-spots' may exist on the surface of globular proteins, i.e. accessible, hydrophilic and 'mobile' features of tertiary structure readily give rise to an immune response. Small molecules (e.g. cyclic

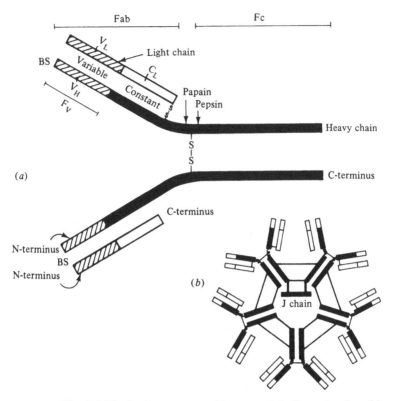

Fig. 2.1 The basic structure of immunoglobulin molecules: (*a*) shows the heavy and light chains held together by disulphide bridges with two binding sites (BS) formed from the highly variable N-terminal region of one heavy and one light chain. The class of the immunoglobulin (IgA, IgD, IgE, IgG, IgM) is determined by the class of the heavy chain. Light chains are either kappa (ϰ) or lambda (λ). Proteolytic cleavage by the enzyme papain produces an Fc (c for crystallizable) fragment made up of two C-terminal heavy chain fragments covalently joined by a disulphide bridge, and two Fab fragments each consisting of a light chain and the remaining N-terminal portion of a heavy chain. Cleavage with pepsin produces Fc fragments and leaves the two Fab fragments covalently linked as an F(ab) fragment. (*b*) IgM is a pentamer of immunoglobulin molecules covalently linked by sulphydryl bridges, and by a single peptide known as the J-chain.

nucleotides or drugs) are not normally antigenic unless injected coupled to larger protein carrier molecules. The small molecule is then called a 'hapten'.

The immune system, when confronted by a new antigen, responds by producing both a 'humoral' defence (antibodies free in solution) and a

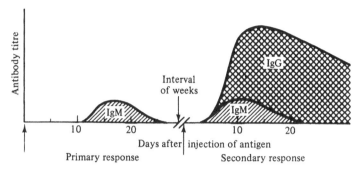

Fig. 2.2. The primary and secondary immune responses. Injection of an antigen at the first arrow leads to the appearance of antibodies after about 10 days. These are mostly of the IgM class. (Contrast this with the appearance of the acute phase protein, CRP, which appears within hours and then disappears within days.) Further injection after an interval of weeks leads to a secondary immune response in which antibodies appear earlier and are now predominantly of the IgG class.

'cell-mediated' defence (sensitized cells which can engulf and remove the injected antigen). The new antigen binds to the surface of small lymphocytes, which then divide and give rise to memory cells and to cells which are responsible for humoral and cell-mediated immunity. T-lymphocytes (thymus-derived or thymus-dependent lymphocytes) are responsible for cell-mediated immunity; B-lymphocytes (bursa-dependent lymphocytes) are responsible for the synthesis of circulating antibody. (The 'bursa' after which B-cells are named is the Bursa of Fabricius found in chickens. The exact equivalent of this organ in Man is uncertain but the population of B-lymphocytes in Man appears to be equivalent to those in the chicken.) Correct B-lymphocyte functioning requires the cooperation of T-lymphocytes.

Following injection of a new antigen into an animal, the primary response in terms of humoral antibody is the production of IgM. Upon a repeated injection, the secondary response produces a marked and prolonged rise of IgG (Fig. 2.2). Repeated injection of a pure antigen into an animal produces an antiserum, i.e. a serum containing a higher level (when compared to the unimmunized animal) of especially IgG immunoglobulin, a high proportion of which (up to 10%) recognizes epitopes (antigenic determinants) on the original injected antigen. In practice, it may take several monthly injections of the antigen (usually emulsified in an oily mixture called Freund's adjuvant which contains killed tubercle bacilli to stimulate the immune system) to produce such a polyclonal antiserum with a high 'titre'. The 'titre' of an antiserum is an arbitrary dilution at which the presence of the antibody can still be detected and its exact value depends on the detection system used. High-titre polyclonal antisera used in some radioimmunoassays for polypeptide hormones (see later) are diluted several hundred thousand times before use in such assays.

Antibodies as specific measuring reagents
The measurement of specific proteins, e.g. the serum proteins discussed above, of polypeptide and other hormones, and of drugs in the circulation of patients is a major preoccupation of hospital laboratories – millions of assays are done in the UK alone each year – and is also one of the main applications to which the specificity and detection sensitivity of antibodies are put (the other being in immunohistochemical techniques, see Chapter 3). Table 2.1 shows the range of concentrations of substances in serum which antibody techniques are used to measure. Serum proteins are present at concentrations varying from about $50\,\mathrm{g\,l^{-1}}$ for serum albumin to $5\,\mu\mathrm{g\,l^{-1}}$ for IgE, while other proteins (e.g. hormones which are en route from an endocrine gland to a target organ and are being carried in serum as 'passenger' proteins) can be present in concentrations as low as $5\,\mathrm{ng\,l^{-1}}$. Antibodies can be used to measure individual proteins in this complex mixture and over this 10^{10} concentration range. This is possible by using two basic types of assay techniques:

1. Immunoprecipitation techniques. These are used for proteins present in serum at concentrations of about $10\,\mathrm{mg\,l^{-1}}$ and above and depend on the formation of a visible macromolecular 'lattice' (immunoprecipitate) between the antigen and the antibody.
2. Competitive binding assays (e.g. radioimmunoassay) where the protein is present in concentrations between $1\,\mathrm{ng\,l^{-1}}$ and $1\,\mathrm{mg\,l^{-1}}$. Here the antiserum is used at high dilution and the antigen has to be labelled.

Immunoprecipitation techniques
Immunoprecipitates, i.e. visible macromolecular lattices of antigen (in this case, a serum protein) and antibody form only at suitable relative concentrations, i.e. at 'equivalence' (Fig. 2.3). Insufficient antigen means that the second binding site of each antibody is unoccupied; if there is too much antigen, cross-linking cannot occur because of insuffi-

Table 2.1 *The concentration range of biological molecules in serum measured by antibody techniques*
Selected examples only of each concentration range are given. β-hCG human chorionic gonadotrophin, ACTH = adrenocorticotrophin, PTH parathyroid hormone, TSH thyroid-stimulating hormone.

Molecule	Molecular weight (kDa)	Molar concentration	Method of measurement
Albumin	66	6×10^{-5}	Dye binding
IgG	160	6×10^{-5}	Radial
Fibrinogen	340	1.2×10^{-5}	immunodiffusion
IgM	950	1.6×10^{-6}	Turbidimetry and
α_1-antitrypsin	87	1.9×10^{-5}	nephelometry
Plasminogen	87	3.4×10^{-6}	Immunoelectro-
Caeruloplasmin	151	2.6×10^{-6}	phoresis
β_2-microglobulin	11	1.7×10^{-7}	
Thyroxine-binding globulin	57	2.1×10^{-7}	Nephelometry
C-reactive protein	132	6.1×10^{-8}	Turbidimetry
Ferritin	500–900	2×10^{-8}	
βhCG	14	2.9×10^{-8}	
			Radioimmunoassay (RIA) IRMA
Insulin	5	1×10^{-10}	
Growth hormone	20	5×10^{-10}	
ACTH	4.68	1.1×10^{-11}	
PTH	10	3.0×10^{-12}	
TSH	28	7.1×10^{-12}	

cient antibody (Fig. 2.3). Note that the protein (antigen) must have more than one antigenic site (or 'epitope') for cross-linking of several molecules to occur, i.e. lattice formation cannot occur with low molecular weight haptens which are too small to carry more than one epitope. Note also that a mixture of different antibodies recognizing different epitopes on the antigen is necessary: a single antibody with two valencies of identical specificity can 'cross-link' only two antigens (each with the respective epitope repeated once on its surface), and further cross-linking leading to lattice formation is not possible. The formation of an

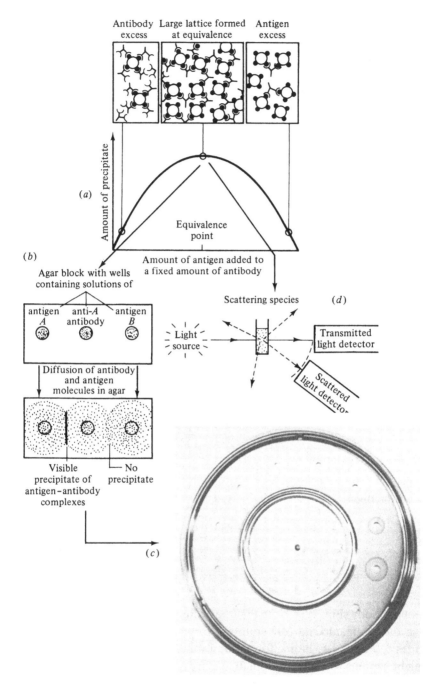

Antibody   Large lattice formed   Antigen
excess   at equivalence   excess

(a)

Amount of precipitate

Equivalence
point

(b)

Amount of antigen added to
a fixed amount of antibody

Agar block with wells
containing solutions of

antigen   anti-A   antigen
A   antibody   B

Diffusion of antibody
and antigen
molecules in agar

Visible   No
precipitate of   precipitate
antigen–antibody
complexes

(c)

Scattering species   (d)

Light
source

Transmitted
light detector

Scattered
light detector

Fig. 2.3. Formation and detection of immunoprecipitates between antigens and antibodies. The formation of a lattice at the equivalence point shown in (a) can be observed as an immunoprecipitate in agar gels as in (b). An actual Mancini plate used in laboratory measurements is shown in (c). Wells 1–4 in the

immunoprecipitate can be observed by radial immunodiffusion in agar plates (Fig. 2.3(*b*)). The distance of the immunoprecipitate from the central well is directly proportional to the amount of antigen (protein) in the sample placed in the well. This technique (Mancini plates, Fig. 2.3) takes 2–3 days and has been superseded by 'rocket' electrophoresis. A variant of this is crossed immunoelectrophoresis (Fig. 2.4) in which serum samples are electrophoresed into antibody-containing agar after prior separation in a first-dimension electrophoresis (Fig. 2.4). These techniques take 1–2 h. A more rapid alternative is to observe the formation of the immunoprecipitate in free solution by nephelometry (measuring increasing light scattering) or by 'kinetic immunoturbidimetry' (measuring increasing absorbance) (see Fig. 2.3). The most rapid available technique at present can measure 20 samples simultaneously in 1–2 min utilizing the centrifugal fast analyser.

Competitive binding assays

The study of patients with endocrine disease was revolutionized by the development in the late 1950s and early 1960s of the technique of radioimmunoassay (RIA), because this procedure made it possible for the first time to measure the levels of peptide hormones in patients with these disorders. The immunoprecipitation techniques described earlier are not suitable for the measurement of peptide hormones at levels down to $5\,\mathrm{ng\,l^{-1}}$ because the antigen (the hormone) is so dilute that a visible lattice cannot form. Antigens at these low levels in serum are instead measured by a 'competition' binding assay of which RIA is the most prominent example. The specific antibody directed against the hormone (or other protein) is diluted (usually many thousandfold) until it half-maximally binds a defined amount of labelled antigen (usually, but not necessarily, radioactively labelled). 'Cold' or unlabelled antigen is then added to 'compete off' the bound antigen from the limiting amount of specific antibody; the more unlabelled or 'cold' antigen is added, the less labelled or 'hot' antigen can remain attached to the antibody. Provided a procedure is available for separating labelled antigen attached to the antibody from unattached labelled antigen, i.e. the 'bound' from the 'free' fraction, the technique can be used to estimate how much unlabel-

Fig. 2.3. (*cont.*)
plate have increasing amounts of antigen added. The agar surrounding the wells contains specific antibody against the antigen. Alternatively, the formation of the immunoprecipitate can be observed in solution (*d*) by measuring light absorbance or light scattering.

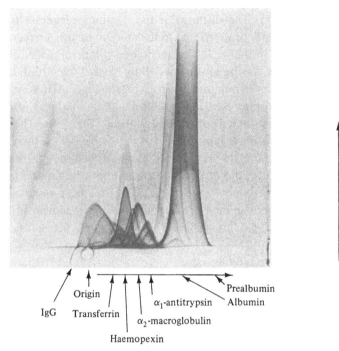

Origin

IgG   Transferrin

Haemopexin

α_1-antitrypsin

α_2-macroglobulin

Prealbumin

Albumin

Fig. 2.4. Crossed immunoelectrophoresis. Human serum proteins have been initially electrophoresed in the direction shown by the horizontal arrow and then electrophoresed in the direction shown by the vertical arrow. In the second dimension, the proteins, initially separated by the first electrophoresis, migrate into an agar gel containing a mixed antiserum raised against whole human serum. At equivalence, they produce 'flames' or rocket-shaped immunoprecipitates. The areas under the peaks are proportional to the amount of the individual protein in the original serum sample. Some of the more prominent serum proteins are labelled.

led antigen must have been present in the original mixture. The RIA technique is illustrated in Fig. 2.5. Note that the 'standard curve' is in a downward direction, i.e. the more unlabelled antigen is present, the less labelled antigen appears in the 'bound' fraction. The RIA technique is still widely used in the assay of peptide hormones, steroid hormones, drugs, and other substances in body fluids and tens of millions of measurements are carried out using this technique annually. The label attached to the antigen does not necessarily have to be radioactive (although this is convenient and easy to quantitate) and other non-

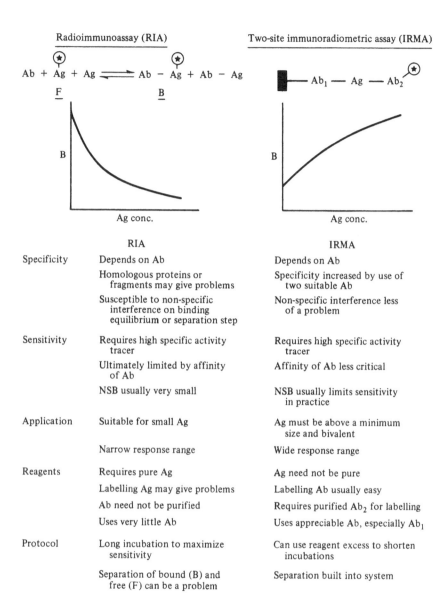

	RIA	IRMA
Specificity	Depends on Ab	Depends on Ab
	Homologous proteins or fragments may give problems	Specificity increased by use of two suitable Ab
	Susceptible to non-specific interference on binding equilibrium or separation step	Non-specific interference less of a problem
Sensitivity	Requires high specific activity tracer	Requires high specific activity tracer
	Ultimately limited by affinity of Ab	Affinity of Ab less critical
	NSB usually very small	NSB usually limits sensitivity in practice
Application	Suitable for small Ag	Ag must be above a minimum size and bivalent
	Narrow response range	Wide response range
Reagents	Requires pure Ag	Ag need not be pure
	Labelling Ag may give problems	Labelling Ab usually easy
	Ab need not be purified	Requires purified Ab_2 for labelling
	Uses very little Ab	Uses appreciable Ab, especially Ab_1
Protocol	Long incubation to maximize sensitivity	Can use reagent excess to shorten incubations
	Separation of bound (B) and free (F) can be a problem	Separation built into system

Fig. 2.5. The basic principles of radioimmunoassay and immunoradiometric assay. The standard curve produced by each technique is shown and the disadvantages and advantages of each are summarized. Note that RIA takes place in free solution, while at least one antibody in an IRMA has to be attached to a solid support, depicted as ▌—. Ab–antibody, Ag–antigen, NSB– non-specific binding, i.e. binding of radioactivity in the assay system in the absence of antigen or antibody.

isotopic labels, e.g. enzymes, fluorescent or chemiluminescent compounds, viruses, red cells, etc. are becoming more frequently used.

One of the inherent disadvantages of the RIA procedure is that the
technique usually involves lengthy (e.g. overnight) incubations and may
take 1 or 2 days to produce a measured value. Furthermore, especially in
the field of peptide hormone measurement, fragments of the hormone
(possibly biologically inactive) as well as the intact (biologically active)
hormone may cross-react. Some antigens (particularly proteins of large
molecular weight) have been found to suffer damage on labelling with
radioactivity or other label and such damage may make the antigen
immunologically unrecognizable by the antibody.

A different approach (again dating from the 1960s) was to use a
labelled antibody rather than a labelled antigen in a procedure known as
an immunoradiometric assay or 'IRMA'. Here the radioactively labelled
antibody is added in excess, all the unlabelled antigen is bound and then
the antigen-labelled antibody complex is separated from free labelled
antibody by using, for instance, a second antibody attached to a solid
support (a 'two-site' IRMA) (see Fig. 2.5). While labelled antibody
assays were useful in some circumstances (e.g. where the antigen in
question was refractory to labelling) and could solve some specificity
problems (e.g. by using two different antisera specific for different parts
of a protein molecule, so that only an intact peptide hormone could be
recognized in the assay rather than its fragments), until recently, labelled
antibody assays represented only a few per cent of the measurements
made by immunoassay in hospital laboratories. This was because the
technique was cumbersome and wasteful of both antigen and antiserum.
Antigen was needed in large amounts to affinity-purify the antibodies for
radioactive labelling: remember that only about 10 % of the IgG molecules in even a high-titre antiserum are specific for the antigen injected,
so these specific antibodies need to be purified before labelling, otherwise
90 % of the antibodies added to the assay do not even recognize the
antigen. Large amounts of antiserum were needed, too, to supply the
purified labelled antibodies. These problems have largely disappeared in
the wake of a major new development, the ability to make 'monoclonal'
antibodies.

Monoclonal antibodies

One result of the injection of an antigen into an animal is that the immune
system produces clones of plasma cells, each secreting one single
individual immunoglobulin (of whatever class) with one single specificity,
different clones of plasma cells contributing to the overall 'polyclonal'

immune response. There is a not-uncommon human disease, known as 'multiple myeloma', in which one individual clone of plasma cells becomes malignant, proliferates uncontrollably and secretes one individual immunoglobulin in large quantities. This disease results in multiple plasma cell tumours in the bone marrow and leads to 'punched out' holes in bones which are visible on X-ray. The individual immunoglobulin synthesized (which is 'irrelevant' and does not recognize a particular antigen) appears in the circulation as a single discrete band known as a 'paraprotein' or 'M-band' (M for monoclonal, not for IgM). The level of the M-band or paraprotein correlates with the mass of malignant plasma cells in the patient's body, i.e. the immunoglobulin acts as a 'tumour marker' and the success or otherwise of treatment can be assessed by following the level of the paraprotein in the patient's serum. It is important clinically to be able to class the paraprotein as an IgG, an IgA or an IgM immunoglobulin (IgD and IgE myelomas are very rare), and to define the light chains as $\varkappa$ or λ. This is done by immuno-electrophoresis. Since the production of the individual monoclonal immunoglobulin is the result of a malignant and therefore uncontrolled process, light chains are often 'over-produced'. These are filtered out by the kidney and appear in the urine as 'Bence-Jones protein' (named after Henry Bence-Jones who noticed in 1845 their unusual property of precipitating out at 50 °C and then re-dissolving at 100 °C, reprecipitating upon cooling). Analysis of serum and urine from a patient with multiple myeloma is shown in Fig. 2.6.

In 1975, Kohler and Milstein developed a revolutionary new technique enabling individual clones of antibody-producing cells to be grown in tissue culture. The antibodies produced are therefore 'monoclonal' in the same way as the paraprotein or M-band in the serum of a patient with multiple myeloma, except that now the specificity of the antibody can be predetermined. In addition, the antibody is 'immortal' and in unlimited supply since the cell line producing it can be maintained indefinitely in culture. Techniques are now becoming available which were not possible with conventional polyclonal antisera.

Producing monoclonal antibodies

After repeated injection of an antigen into an animal, immune cells which are capable of secreting antibody against that antigen accumulate in the spleen. These cells will now grow and divide in tissue culture. Malignant antibody-producing cells (myeloma cells, as seen in a patient with multiple myeloma) will grow and divide in tissue culture and will secrete antibody, although the antibody produced is 'useless' and not directed

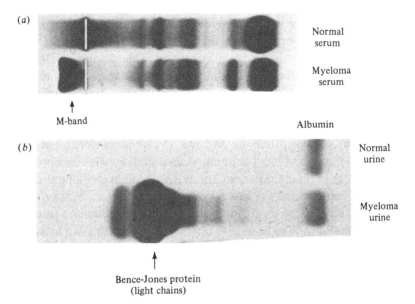

(a)

Normal
serum

Myeloma
serum

M-band

Albumin

(b)

Normal
urine

Myeloma
urine

Bence-Jones protein
(light chains)

Fig. 2.6. Analysis of serum and urine from a patient with multiple myeloma. (*a*) A normal serum electrophoretic strip is shown above, with the patient's pattern below. Note the presence of the monoclonal immunoglobulin (the M-band) present in $30\text{--}40\,\mathrm{g}\,\mathrm{l}^{-1}$. Note also that there is little normal immunoglobulin, i.e. the normal staining produced by IgG, IgA and IgM (see Fig. 1.10) is greatly reduced. This is because the malignant plasma cells producing the M-band have crowded out normal plasma cells. The patient is therefore 'immunosuppressed' and prone to infections. (*b*) Analysis of the proteins in the patient's urine. Normal urine contains very little protein (about $0.15\,\mathrm{g}$ per 24 hours' urine production) and therefore is concentrated before being electrophoresed as with serum protein analysis. Note that in the normal urine there is only a trace of serum albumin present, while in the multiple myeloma patient there are large quantities of Bence-Jones protein representing light chains from the malignant overproduction of a monoclonal immunoglobulin.

against any known antigen. By fusing the spleen cells from the immunized animal (which have been programmed by the injection of antigen to produce 'useful' antibodies against that antigen) with the malignant myeloma cells, a hybrid cell is produced which can grow in culture and will secrete 'sensible' antibody.

The procedure (and the contrast with conventional antiserum production) is shown in Fig. 2.7. In practice, a mouse myeloma cell line is used, so the programmed spleen cells are also taken from a mouse. The mouse myeloma cell has been specially selected so that it lacks the enzyme hypoxanthine guanine phosphoribosyl transferase (HGPRT). This is because the process in which spleen cells are fused with malignant myeloma cells is not very efficient and only 1 in 10^3 or 1 in 10^5 cells produce viable hybrids. As the myeloma cells grow very rapidly in culture (in contrast to the spleen cells, which do not grow and divide at all), the small proportion of hybrid cells produced is rapidly outgrown by the malignant myeloma cells and they die. HGPRT is an enzyme in the salvage pathway of purine biosynthesis which reclaims hypoxanthine and guanine from AMP and GMP breakdown and forms inosine and guanosine without having to resynthesize the entire purine ring. As the malignant myeloma cells do not possess HGPRT, they cannot use the salvage pathway and are forced to use the *de novo* synthesis route. This can be blocked by aminopterin. The hybrid cells, however, do possess HGPRT (derived from the normal spleen cells which form one source of the hybrid) and therefore can survive in the presence of aminopterin by using the salvage pathway. Thus, by growing the mixture of hybrid cells and malignant myeloma cells in a medium containing hypoxanthine, aminopterin and thymidine, called HAT medium (the thymidine is present because aminopterin also blocks pyrimidine synthesis), the hybrid cells are selectively favoured while the myeloma cells die. By repeatedly diluting and regrowing the hybrid cells at limiting dilution (cloning), each individual cell eventually forms a single clone secreting one monoclonal antibody.

Monoclonal antibodies as measuring reagents
Monoclonal antibody techniques have now made possible the production of unlimited amounts of specific antibodies which do not need prior purification before labelling. These antibodies can be used in 'two-site' procedures, which are more specific assays because a particular antigen has to possess two distinct epitopes to be recognized by the assay, e.g. fragments of hormones do not react, only the intact hormone can. Monoclonal antibodies are also making possible much more rapid assays (taking a few hours rather than 2–3 days as with conventional RIA). In addition, newer assays are possible in which pairs of monoclonal antibodies against two different epitopes on the same antigen are used cooperatively (Fig. 2.8(*a*)). Note the difference between the *affinity* of an antibody binding site for its specific epitope (which can be analysed in mass action terms) and the *avidity* of an antibody system for an antigen

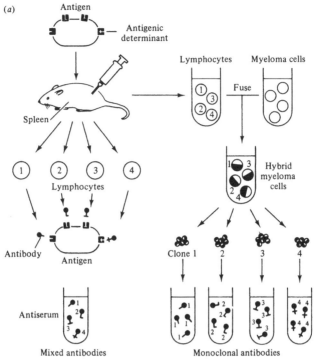

(a)

Antigen

Antigenic determinant

Lymphocytes   Myeloma cells

Fuse

Spleen

Hybrid myeloma cells

Lymphocytes

Clone 1   2   3   4

Antibody

Antigen

Antiserum

Mixed antibodies

Monoclonal antibodies

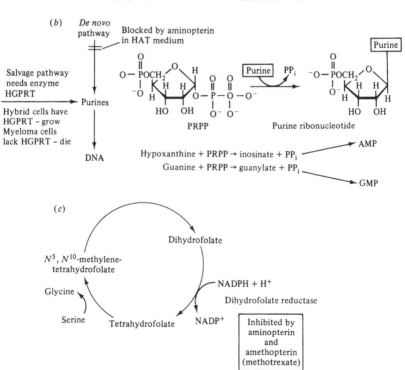

(b)   *De novo* pathway   Blocked by aminopterin in HAT medium

Salvage pathway needs enzyme HGPRT

Hybrid cells have HGPRT – grow
Myeloma cells lack HGPRT – die

Purines

DNA

PRPP

Purine

PP_i

Purine ribonucleotide

Purine

Hypoxanthine + PRPP → inosinate + PP_i ⟶ AMP

Guanine + PRPP → guanylate + PP_i ⟶ GMP

(c)

Dihydrofolate

N^5, N^{10}-methylene-tetrahydrofolate

Glycine

Serine

Tetrahydrofolate

NADPH + H^+

Dihydrofolate reductase

$NADP^+$

Inhibited by aminopterin and amethopterin (methotrexate)

(which depends on antibody multivalency, multiple determinants on the antigen, steric factors, etc.). In Fig. 2.8(*b*), the individual affinities at each binding site (1, 2, 3, 4) may be low, but the overall *avidity* of the monoclonal antibody pair for the antigen is high. As a general rule, the higher the avidity, the more rapid and sensitive the immunoassay.

'Proximal linkage' assays can be designed which do not require the separation of 'bound' from 'free' antigen. This is illustrated in Fig. 2.8(*c*). Two monoclonal antibodies, each directed against different epitopes on an antigen and each covalently coupled to a specific reagent, are bound by the antigen and are thus held in close proximity. By generating a signal which can pass from one antibody to another, e.g. fluorescence quenching or chemiluminescent transfer, assays can be performed in free solution with no separation steps.

Monoclonal antibodies as affinity purification reagents
Several biological macromolecules are clinically important but very difficult or impossible to purify in large quantities. Examples are the antiviral and anticancer agent interferon, and factor VIII necessary to stop haemophiliacs from bleeding. Most monoclonal antibodies are of low affinity (reflecting the population of antibodies in the intact mouse) and therefore are quite suitable for extracting the desired substance from impure mixtures. The great advantage of monoclonal antibodies (apart from abundant supply, hence large affinity columns can be made) is that the antibodies can be raised against *impure* preparations of the desired substance and then screened to find which antibodies are specific. Cloned cells producing these can then be grown in bulk and the resultant monoclonal antibodies used to prepare large quantities of the pure

Fig. 2.7. The production of monoclonal antibodies. (*a*) Two methods of producing antibodies are illustrated: the conventional method (left) consists of repeated injections followed by bleeding the animal to produce a polyclonal antiserum; in the monoclonal antibody method (right), the spleen is removed after one or two injections and spleen cells are fused with myeloma cells to produce individual monoclonal antibodies in culture. Selection of viable hybrid cells is by the HAT method. The *de novo* purine synthesis route is blocked by aminopterin (methotrexate), which inhibits dihydrofolate reductase (see (*c*)). HGPRT functions in the salvage pathway of purine synthesis as shown in (*b*). Since the myeloma cells lack HGPRT, only myeloma cells which have fused with spleen cells (which contain the enzyme) to form antibody-producing hybridomas will survive.

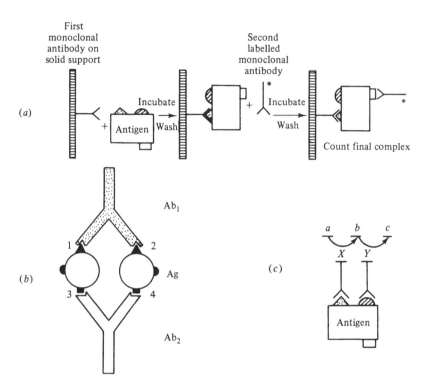

Fig. 2.8. Monoclonal antibodies in immunoassay systems.
(a) A two-site labelled antibody assay using two different
monoclonal antibodies directed against two different epitopes on an
antigen. Such assays are very specific as the antigen must have both
epitopes to react (e.g. biologically inactive fragments of hormones
will not react) and can be rapid compared to RIA systems since
large amounts of antibody can be used (monoclonal antibodies are
available in potentially limitless supply).
(b) The formation of cyclic complexes between two antibodies and
two antigens. The individual affinity at each site may be weak, but
the overall avidity of all four sites considered together may be very
high. The assays can therefore be very sensitive. Similar complexes
may form in polyclonal antibody systems.
(c) A proximal linkage assay. Two different monoclonal antibodies
are covalently linked to signalling molecules X and Y. When these
antibodies are bound to their specific antigen, X and Y are brought
into sufficiently close proximity for a signal to pass between them.
This signal can be, for instance, chemiluminescent or fluorescent or
even 'enzyme channelling'. Hence, a could be a particular excitatory
wavelength of light or a substrate for molecule X. b could be a
wavelength of light emitted by X and excitatory to Y, or the product
of enzyme activity by X and a substrate for Y. Detection of c can be
a measure of how much of the antigen is present. Such assays can be
very rapid and are said to be 'homogeneous', since no separation
step of bound antigen is required (cf. RIA and IRMA). Reproduced
with permission from *Trends in Biochemical Sciences*, Elsevier,
Cambridge.

substance. A protocol for the preparation of interferon is shown in Fig. 2.9.

Monoclonal antibodies as labelling reagents in cell sorting

Monoclonal antibodies can recognize different antigens on the surface of, for example, different white cells (Fig. 2.10). Fluorescent labels of different sorts can be attached to these antibodies (X and Y in Fig. 2.10) and a fluorescence-activated cell sorter (FACS) can be used to isolate individual cell types. As the fluorescently labelled cells pass through the sorter, the stream of liquid is broken up into drops. According to the reading on the fluorescence detector, a charge can be applied to the deflector plates and the droplet deflected to left or right. The forward scatter detector measures light scattering, which is a rough guide to the size of the cell. Thus, cells can be selected for size, and intensity of fluorescent labelling, i.e. content of specific antigen on the surface. Such techniques have been used to isolate foetal cells from maternal blood, i.e. to obtain samples of foetal cells without puncturing the uterus.

Monoclonal antibodies as targeting reagents for cytotoxic therapy

The unique specificity of monoclonal antibodies makes possible attempts to kill specific cell types within a mixture of cells, either in cell populations extracted from the patient (e.g. blood cells) or in the whole patient. While it is not yet clear whether antigens are present on the surface of malignant cells which are completely absent from normal cells (so called 'tumour-specific' antigens), it is clear that monoclonal antibodies can distinguish different normal cell types very easily. Several toxic substances have been coupled to monoclonal antibodies to form 'magic bullets', i.e. any one cell type is recognized by the monoclonal antibody and attacked by the toxin. An example is shown in Fig. 2.11. Several toxins exist which consist of two subunits, neither of which in isolation is toxic to cells. In the case of such toxins as abrin and ricin, one molecule of each is enough to kill a single cell. The B subunit of ricin (shown in Fig. 2.11(a) binds to galactose residues on cell-surface glycoproteins; the A subunit is then released and penetrates the cell where it inactivates the 60S ribosome. By substituting a monoclonal antibody towards a cell-type specific antigen for the B subunit (Fig. 2.11(b)), the A subunit can be carried to specific cell types. This kind of cytotoxic therapy has been used to kill lymphocytes in bone marrow preparations which are about to be grafted into recipients, thus reducing the risk of 'graft-versus-host' reactions.

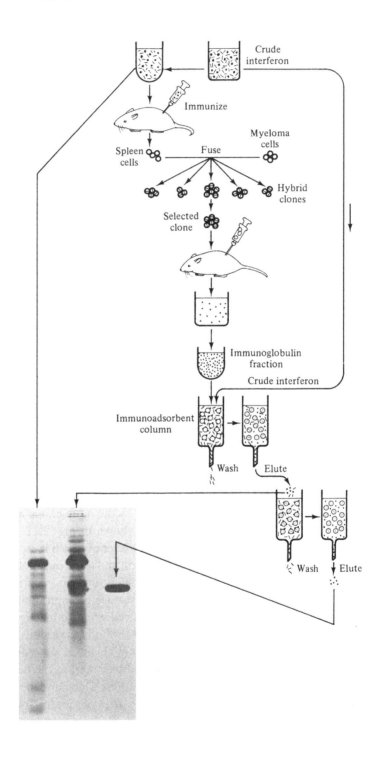

Monoclonal antibodies to drugs as therapeutic reagents

Monoclonal antibodies can specifically recognize and bind drugs, and some drugs are more rapidly cleared from the circulation when bound to antibody. An example is shown in Fig. 2.12 of a patient who took an overdose of digitoxin, a drug used in cases of heart failure. Overdose with such a drug can produce a fatal cardiac arrhythmia. Fab fragments, i.e. fragments of monoclonal antibody specific for digitoxin produced from the intact immunoglobulin molecule by papain digestion, were injected into the patient's circulation. The fragments rapidly bound the drug and the antibody–drug conjugate was rapidly removed by the reticulo-endothelial system. Since the affinity of the antibody for the drug may be higher than that of the physiological receptor for the drug, toxic effects may be rapidly reversed.

Monoclonal antibodies as diagnostic imaging reagents

The pathological diagnosis of many diseases is achieved by consideration of the macroscopic appearance of tissue specimens followed by careful microscopic examination. The advent of monoclonal antibodies (because of their purity and availability in unlimited quantities) has allowed the more widespread use of immunological techniques to identify specific antigens in tissue sections and thus aid diagnosis. Usually, such techniques identify sites of monoclonal antibody binding by visualization with a second anti-immunoglobulin antibody labelled with a fluorescent or enzymic probe.

Immunohistological techniques have shown that, while it has been difficult to demonstrate any truly tumour-specific antigen, there are many antigens which because of considerable quantitative differences between normal and malignant tissue can act as tumour markers. It has been possible to use radiolabelled monoclonal antibodies to some of these to identify tumours in living patients (Fig. 2.13). This diagnostic imaging technique provides diagnostic information which may be particularly

Fig. 2.9. An example of the use of monoclonal antibodies in the immunopurification of valuable biological macromolecules. Monoclonal antibodies are raised to a crude preparation of interferon. Those reacting to interferon (detected using very small amounts of the purified substance) are grown in bulk and used as an affinity column (see p. 17). Repeated passage of the crude interferon preparation through such immunoaffinity columns results in the bulk isolation of pure interferon.

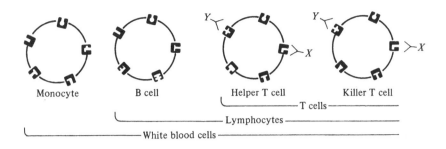

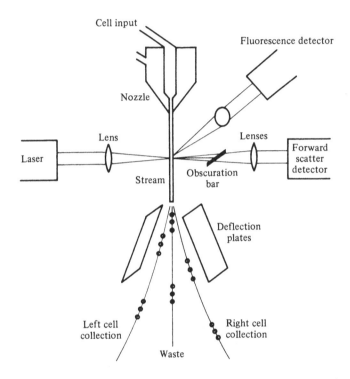

Fig. 2.10. Cell-sorting using fluorescently tagged monoclonal antibodies. Different types of white blood cell can be isolated by tagging with fluorescently labelled monoclonal antibodues, see text for further details.

helpful in detecting unsuspected metastases, thus influencing the choice of surgical operation and subsequent patient management.

Monoclonal antibodies as anti-idiotype antibodies

Each individual antibody molecule that arises after the injection of an antigen possesses a binding site which recognizes a particular epitope on that antigen. This binding site is formed from the hypervariable regions of the heavy and light chains of the antibody and the structure of this region

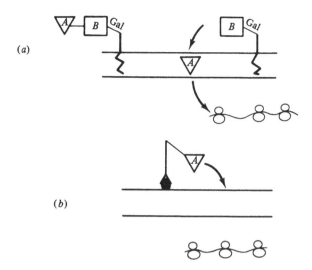

Fig. 2.11. Monoclonal antibodies targeting cytotoxic therapy. (*a*) The *A* subunit of the toxin ricin disrupts translation on cytoplasmic ribosomes. (*b*) Monoclonal antibodies directed against antigens on specific cell types can be used to target the toxin to particular cells. See text for further explanation.

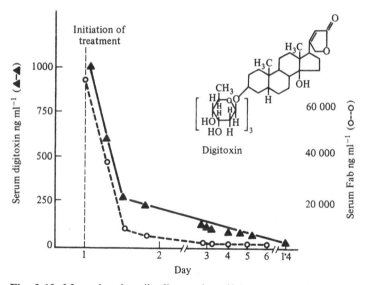

Fig. 2.12. Monoclonal antibodies as detoxifying agents. The patient has taken an overdose of the cardioactive drug digitoxin (structure shown). Fab fragments from mouse monoclonal antibodies against the drug are injected. These bind the drug in the circulation and inactivate it. The complex is then removed via the kidneys. (Fab fragments are used because they are less antigenic than the whole molecule also containing the Fc fragment.)

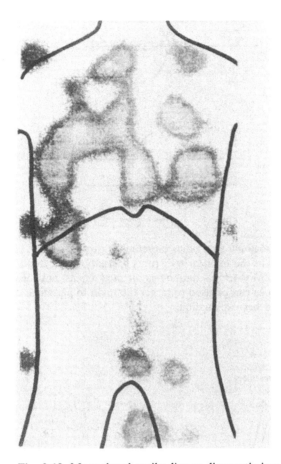

Fig. 2.13. Monoclonal antibodies as diagnostic imaging reagents. Isotopic 'scan' using a radiolabelled monoclonal antibody directed against an antigen on the surface of a tumour (a colorectal carcinoma). There is uptake of label by tumour metastases (secondary deposits of tumour) as well as by the primary colorectal carcinoma. Courtesy of Professor K. Sikora.

is not only unique but also (because it is unique) 'foreign' to the host animal. This individually unique structure on each antibody is known as an 'idiotypic determinant' or 'idiotype', and antibodies produced against this internally generated 'foreign' structure are termed 'anti-idiotype' antibodies. N. Jerne proposed that the immune system consisted of a network of antibodies, each directed against each other's idiotypes. Introducing a foreign antigen causes overproduction of one type of antibody, which stimulates production of the respective anti-idiotype antibodies, etc., thus limiting the interaction of the first antibody with the antigen. According to 'network theory', the encounter of a foreign antigen by the system is analogous to ripples spreading across the surface of a pond.

Because the epitope on the surface of an antigen and the binding site of an anti-idiotype antibody both interact with the idiotypic region of the antibody directed against the antigenic epitope, they must have structural similarities. This suggests that at least some anti-idiotype antibodies have variable regions which are internal images of the relevant antigen and which closely resemble regions of the antigenic structure. This realization has led to two promising exploitations of anti-idiotype antibodies, first in the purification of receptors and secondly as novel vaccinating agents.

There is much interest in the structure and function of receptors, either for hormones (see Chapter 5) or for neurotransmitters. These proteins, however, are usually present at low levels in tissue and, furthermore, because of their hydrophobic, membrane-bound environments, are often technically difficult to purify in sufficient quantities for protein studies or for antiserum production. A new approach has been to use monoclonal anti-idiotype antibodies. Monoclonal anti-idiotype antibodies are raised against monoclonal anti-hormone antibodies since the idiotypic determinants on some of these will resemble the structure of the receptor. Thus, anti-receptor antibodies can be produced against the receptor without the receptor being used as antigen. These antibodies may be used for subsequent immunoaffinity purification of the receptor and to probe receptor structure and function.

Most vaccines contain the pathogenic organism (or components of the pathogen) against which protection is desired, the vaccine having been treated in some way to render the pathogen harmless. Other vaccines are 'live' and consist of alternated strains of the pathogen which are non-pathogenic but produce the appropriate immune response. Because side-effects are possible from various components of the vaccine (or even from pathogenic material itself), much interest is being shown in immunizing with pure non-pathogenic material which should be free from such side-effects. Anti-idiotypic antibodies (especially monoclonal) offer this possibility, since antibodies can be raised in animals against components of the pathogen, anti-idiotype antibodies can be raised against these primary antibodies, and these anti-idiotype antibodies can be used as direct vaccines to produce the required immune response without risk of side-effects. A further possibility for the future is that anti-idiotype antibodies (raised against antibodies to tumour-specific antigens on the surface of malignant cells) might eventually be used to develop cancer vaccines.

Genetically engineered monoclonal antibodies

A major problem associated with the use of antibodies as therapeutic and diagnostic reagents is that, as foreign proteins, they themselves induce an immune response resulting in neutralization of their function. This is true of mouse monoclonal antibodies administered to patients, but the problem would be much less serious if human antibodies could be used. Unfortunately, there is no simple generalized way of obtaining human monoclonal antibodies of the right specificity. However, the techniques of genetic engineering have been used to obtain antibodies in which the antigen-binding site is defined by protein sequence from a mouse monoclonal antibody of the right specificity and the rest of the molecule is human. This has been possible because a number of vector systems have been developed that allow the expression of antibody light chain and heavy chain genes in cells transfected with them. While a detailed discussion of such approaches is beyond the scope of this book, basic genetic engineering techniques are described in Chapter 7. It is hoped that, by constructing chimaeric antibodies with mouse hypervariable regions embedded in a human variable region framework, antibodies can be produced that will be sufficiently 'self-like' to become integrated into the patient's immune network without causing a further immune response.

A further use of genetically engineered antibodies is to produce molecules in which the Fc portion of the antibody is replaced by an unrelated protein. This has been achieved experimentally to produce a molecule containing an antigen binding site and an enzyme activity. The potential applications of this technology for such purposes as enzyme-linked immunoassay and targeting chemotherapeutic drugs offer exciting prospects for the future.

3

Tissue-specific proteins

Identification of tissue-specific proteins

Although the human genome probably contains enough single-copy structural genes to code for between 30 000 and 50 000 proteins, only a small percentage of these are actively expressed in any particular cell type at any one time. Some proteins, e.g. the enzymes of ubiquitous metabolic pathways (such as glycolysis) or common 'regulatory' proteins such as calmodulin, are apparently present in virtually all cell types and fulfil 'housekeeping' functions. Others, e.g. those forming cytoskeletal elements or enzymes synthesizing special metabolites, appear to be specific to particular cell types. Clinically, biochemistry exploits patterns of tissue-specific proteins as a non-invasive way of estimating damage to particular organs within the body. This is done by measuring the levels of individual tissue-specific proteins (often enzymes) in the circulation on the premise that the greater the damage to the tissue of origin, the greater the amount of tissue-specific protein released into the bloodstream.

There are several ways of establishing the tissue-specificity of a particular protein. If the protein is an enzyme, direct measurement of catalytic activity may show higher levels in one organ than another, e.g. creatine kinase occurs in large amounts in skeletal muscle and can be measured in serum to diagnose diseases such as muscular dystrophy, but it is virtually absent from liver. (Note that high levels of catalytic activity in more than one human organ may still be diagnostically useful if an isoenzyme system is present – see p. 65.) If the protein is not an enzyme but an antibody is available, human tissues can be surveyed for the presence of the protein by some form of immunoassay (see Chapter 2). New tissue-specific proteins can also be found by analysing the spectrum of proteins within a tissue using high resolution techniques such as two-dimensional electrophoresis. By comparing the patterns from different organs, proteins specific to particular organs can be selected. This is illustrated in Fig. 3.1, which shows a 'map' of the major soluble (i.e.

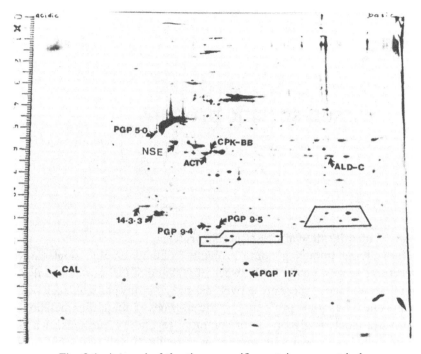

Fig. 3.1. A 'map' of the tissue-specific proteins present in human brain. Soluble (cytoplasmic) proteins present in human brain are shown analysed by high-resolution, two-dimensional polyacrylamide gel electrophoresis (see p. 21). Proteins such as calmodulin (see p. 152) and actin (see p. 59) are present in all human organs, as are the proteins (of unknown function) enclosed in the black lines. Such constant 'constellations' of proteins are useful to orientate positions on the 'map'. Proteins labelled NSE, CPK-BB, and ALD-C represent the subunits of brain-specific isoenzymes of enolase, creatine kinase and aldolase, respectively. Other proteins (14-3-3 and those pre-fixed with 'PGP') are abundant human brain-specific proteins of unknown function but whose detection in serum and cerebrospinal fluid may be diagnostically useful in various neurological disorders.

cytoplasmic) proteins in human brain. Note that human organs are not homogeneous collections of one cell type but contain blood vessels, nerves, fibrous tissue, etc. The brain is an extreme example of such heterogeneity, being made up of a complex mixture of neurones and glial (supporting) cells, both of many different individual types. The most useful technique in localizing proteins to individual cell types within a particular tissue is immunohistochemistry, whereby thin sections of a tissue are 'stained' using a specific antibody to the protein of interest. Only cells containing the protein show up. The technique is illustrated in Fig. 3.2.

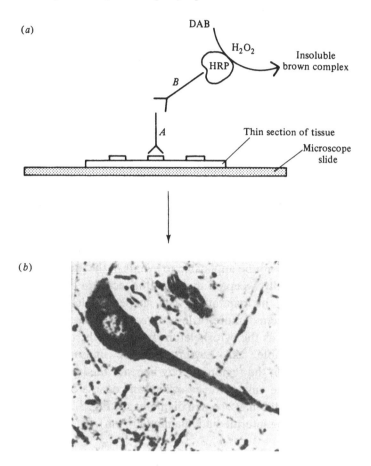

Fig. 3.2. The technique of immunohistochemistry. (*a*) A thin section of tissue is placed on a microscope slide and a specific first antibody (labelled *A*), directed against a protein of interest, is applied to the slide. The antibody binds to those cells containing the protein, and this binding is detected by applying a second antibody (labelled *B*), which is directed against the first antibody (e.g. if antibody *A* was a rabbit antiserum, then *B* would be a sheep-antirabbit-immunoglobulin or a mouse monoclonal antibody against rabbit immunoglobulin). Antibody *B* is coupled to horseradish peroxidase (HRP) which, in the presence of hydrogen peroxide, turns a dye (diaminobenzidine or DAB) into a brown insoluble formazan complex. This is deposited only over the cells containing the initial protein of interest and these appear brown-stained when subsequently viewed down the microscope. (*b*) The photograph shows this technique used on human brain with an antibody against NSE (neurone-specific enolase, see Fig. 3.1). A large neurone is heavily stained with processes of other nerve cells seen criss-crossing the picture.

The cytoskeleton

The cytoplasm of a eukaryotic cell (i.e. the remaining 'soluble' portion of the cell after removal of subcellular organelles including endoplasmic reticulum and ribosomes) consists of a highly concentrated (20–30 %) 'solution' of proteins. This promotes weak interactions between individual proteins which would not be detectable in dilute solution, and also orders the water inside the cell into two phases, i.e. water of hydration, which is tightly bound to the surface of proteins, and free water, which is not. Metabolites may diffuse with varying ease through these two water phases. It is also possible (though not proven) that many metabolic pathways may exist in multienzyme macromolecular complexes held together by the protein interactions possible in highly concentrated solutions. Because of this 'organization' of the cytoplasm, metabolic fluxes and reaction rates predicted from measurements in dilute solution may not be applicable to events in the intact cell.

It is now generally accepted that a major factor in organizing intracellular cytoplasm, in maintaining cell shape and in promoting cell locomotion is a system of more-or-less permanent filamentous structures collectively known as the cytoskeleton. In most cells, the cytoskeleton permeates the whole cell. Erythrocytes are unusual in having a cytoskeleton only in the region directly beneath the plasma membrane (see Fig. 4.22). The eukaryotic cytoskeleton consists of three types of filament: the microtubule (24 nm in diameter see Fig. 3.3), the microfilament (7 nm) and the intermediate filament (10 nm).

Microtubules

Microtubules are seen in virtually all mammalian cells, with the exception of the erythrocyte. All microtubules are formed from α- and β-tubulin, each of 55 kDa. These normally exist as $\alpha\beta$-dimers (110 kDa) and polymerize in a head-to-tail fashion ($\alpha\beta \rightarrow \alpha\beta \rightarrow \alpha\beta$) to form a protofilament. These protofilaments, aligned into 13 parallel rows, form the completed microtubule (Fig. 3.4). Microtubules can be isolated from tissues by repeated chilling of tissue extracts, followed by warming in the presence of GTP and absence of calcium, and then centrifugation. This unusual purification procedure is possible because microtubules de-polymerize into tubulin ($\alpha\beta$-dimers) on cooling and re-polymerize on warming under these conditions. *In vitro*, the polymerization of tubulin into microtubules can be shown to occur by the preferential addition of $\alpha\beta$-dimers to one end (the assembly end) of a preformed microtubule fragment; at the other end of the microtubule (the disassembly end) $\alpha\beta$-dimers are preferentially lost (see Fig. 3.5). Each $\alpha\beta$-tubulin dimer binds

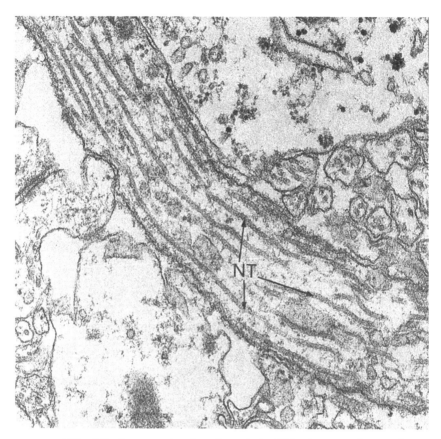

Fig. 3.3. An example of a microtubule. This electron micrograph shows a portion of an axon belonging to a neurone in the human brain. Neurotubules (NT) pass down the axon from the cell body to the synapse. Magnification 60 000×. Courtesy of Professor R. O. Weller.

two molecules of GTP, one of which is hydrolysed to GDP during or shortly after addition to the growing end of the microtubule. *In vitro*, a steady-state can be achieved in which the rate of addition of $\alpha\beta$-dimers to the assembly end of the microtubule is balanced by the loss of $\alpha\beta$-dimers from the disassembly end. An $\alpha\beta$-dimer which has been specifically 'tagged' with radioactive GTP can then be seen to 'move' down the microtubule (Fig. 3.5). It is not clear whether this process – known as 'treadmilling' – actually occurs *in vivo*. Microtubules appear to be able to assemble and disassemble rapidly inside the cell and are important structures in the transport of materials along axons (axonal flow) as seen with neurotubules in neurones, in separating chromosomes at mitosis

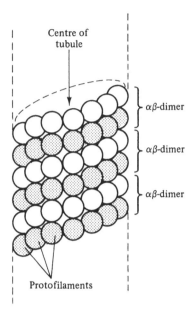

Fig. 3.4. Protofilaments (13) polymerize to form the intact microtubule.

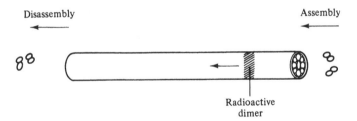

Fig. 3.5. 'Treadmilling' of microtubules. $\alpha\beta$-tubulin dimers are added to the assembly end of a microtubule and removed from the disassembly end. A free $\alpha\beta$-tubulin dimer binds two molecules of GTP, only one of which is removed when the $\alpha\beta$-dimer polymerizes into the tubule at the assembly rod. By using radioactive GTP, the radioactive dimer can be seen to 'migrate' down the tubule.

where microtubules form the mitotic spindle, and as the central components of cilia, e.g. those lining the respiratory tract.

Purification of microtubules from different tissues often shows the presence of other co-purifying proteins known as microtubule-associated proteins or 'MAPS'. It is likely that these proteins, which appear to vary from tissue to tissue, confer unique properties on microtubules polymerized from the same $\alpha\beta$-dimer motif. Several isoelectric variants of α- and β-tubulin are known in different human tissues and the tubulin genes appear to be part of a multi-gene family. Several of the genes thought to code for functional tubulins are now known, however, to be various forms of 'pseudo-genes' (see Chapter 7). In humans, there are probably at least two distinct α-tubulin and three distinct β-tubulin genes which produce functional tubulin mRNAs. While these code for closely related protein molecules, it is not clear how much differential expression of particular tubulin sub-types occurs in different tissues.

Microfilaments
Actin (42 kDa) is present in all eukaryotic cells and is often the most abundant 'cytoplasmic' protein present. Like tubulin, actin genes form a super-family and several very similar forms of actin are known. In muscle cells, at least four different forms of α-actin are present: one for striated skeletal muscle, one for striated cardiac muscle, one for smooth muscle in the walls of blood vessels, and one for smooth muscle in the wall of the gut. These show minor differences in amino acids amounting to between four and six substitutions in a sequence of approximately 400 amino acids. In both muscle and non-muscle cells, β-actin and γ-actin are also present. These differ in about 25 amino acids from striated muscle α-actin.

Monomeric or 'G-actin' can polymerize into 'F-actin' filaments, and this is promoted by Mg^{2+} and by high concentrations of K^+ or Na^+. Actin filaments, which can be shown to 'treadmill' *in vitro* like microtubules, are formed by the addition of G-actin monomers (containing a tightly bound ATP molecule) to the growing end of the filament. Shortly after polymerization, ATP is hydrolysed to ADP and G-actin-ADP monomers are released from the 'disassembly' end of the F-actin polymer. In skeletal muscle actin, 'thin filaments' interact with myosin 'thick filaments' to produce muscle contraction. (Myosin can be regarded as an actin-activated ATPase.) Another example of a permanent microfilament array occurs in the microvilli of intestinal epithelial cells, where bundles of actin filaments perform a structural role in maintaining the shape of the microvilli. All actins (from muscle or non-muscle sources) can co-polymerize and can activate myosin-ATPase. Myosin also occurs in non-muscle mammalian cells but in an unpolymerized form.

Actin-binding proteins
While F-actin and G-actin are in equilibrium at the concentrations of ATP and ions believed to exist within a cell, the formation of F-actin should be highly favoured. Actin is kept in the unpolymerized G-form by

the presence of actin-binding proteins (especially profilin), which bind to monomeric G-actin and prevent its polymerization. Several other proteins are now known which will bind to F-actin in an apparently specific manner. These (apart from myosin found in skeletal muscle) include vinculin and α-actinin which mediate the attachment of actin filaments to cell membranes, fimbrin which cross-links adjacent filaments to produce actin fibres, and various 'capping' proteins which bind to the ends of F-actin filaments and prevent loss or addition of actin monomers. The spectrum of actin-binding proteins varies in different cell types and emphasizes the complexity of the mechanisms involving actin filaments in cell shape and motility.

Intermediate filaments

Under the electron microscope, these cytoskeletal structures appear as filaments of 10 nm diameter and are thus 'intermediate' between actin microfilaments (6 nm) and microtubules (24 nm). Intermediate filaments are rarely straight and often appear as coil-like arrangements, especially around the nucleus. In many cell types, the amount of intermediate filament proteins can equal that of actin. Although intermediate filaments appear microscopically structurally similar in different cell types, antibodies and sequence studies have classified them into five main groups which are summarized in Table 3.1. The appearance of cytoskeletal intermediate filaments under the light microscope in one particular cell type is shown in Fig. 3.6.

Three intermediate filament proteins (vimentin, desmin and glial fibrillary acid protein – see Table 3.1) are single proteins; the neurofilament proteins occur as three polypeptide chains. Human keratins occur in 19 different forms, separable according to two-dimensional gel appearances and sequence data. This family of keratins can be subdivided into Type I (keratins 1–9) and Type II (10–19). These keratins show tissue-specific localizations, e.g. numbers 4 and 13 are most prevalent in the epithelia lining the oesophagus and tongue, while 3 and 12 are specific for the cornea. The five classes of intermediate filament proteins are present in primitive vertebrates and have thus been highly conserved during evolution. The number and position of introns (but not the length of the intron or its sequence) are well preserved in the genes coding for the individual intermediate filament proteins, with the exception of the smallest (63 kDa) neurofilament protein, where the introns are in different positions. This suggests that the neuronal and non-neuronal intermediate filaments may have separated early in evolution and that the keratins, vimentin, desmin and glial fibrillary acidic protein filaments evolved at a later date.

Table 3.1 *Types of intermediate filament (IF) proteins*

Type of IF protein	Molecular weight (kDa)	Numbers of polypeptides	Cell types
Vimentin	54	1	Fibroblasts Chondrocytes Endothelial cells
Desmin	53	1	Skeletal muscle Smooth muscle
Keratins	40–68	19	Epithelial keratinocytes
Glial fibrillary acidic protein (GFAP)	51	1	Astrocytes in brain
Neurofilament proteins	63, 160, 200	3	Neurones in central and peripheral nervous systems

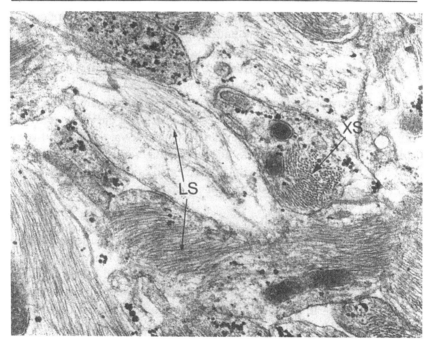

Fig. 3.6. Intermediate filaments in a particular cell type. This electron micrograph shows intermediate filaments in astrocyte processes in human brain. Astrocytes are supporting (glial) cells in the brain and play an important role in the blood–brain barrier. The filaments are composed of glial fibrillary acidic protein and are seen in longitudinal (LS) and cross-section (XS) as are mitochondria. Magnification 60 000×. Compare the diameter of these 'intermediate' filaments with those of neurotubules (Fig. 3.3). Electron micrograph courtesy of Professor R. O. Weller.

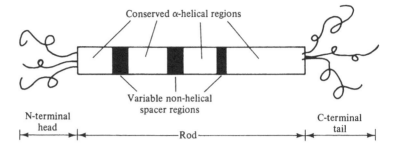

Fig. 3.7. Common structural motifs in intermediate filament proteins. All intermediate filament proteins contain a central core portion of three α-helices unwrapped in a left-handed superhelix, although individual IF proteins have differing 'head' and 'tail' regions.

Despite their sequence differences, individual types of intermediate filaments appear very similar under the electron microscope. This is because each shares a central rod portion which shows a 30–70 % homology with different intermediate filaments. The central helices can wind round each other in a coiled coil arrangement (Fig. 3.7). The variability between individual intermediate filament proteins comes from additions to this central coiled-coil region, a 'head' at the amino terminal end and a 'tail' at the carboxyl terminal end. These vary greatly between different types of intermediate filament, being very short, as with the head regions of keratin 19 or glial fibrillary acidic protein, or very long, as with the tails of the larger neurofilament proteins.

Intermediate filaments and disease

Alzheimer's disease is a common form of senile dementia affecting 1 in 10 of people over the age of 65. A characteristic feature of this disease is the accumulation within nerve cells in the brain of 'neurofibrillary tangles'. These are composed of masses of paired helical filaments, each filament being 100–130 Å in diameter and winding around its partner once about every 650 Å. These paired helical structures are extremely insoluble in agents which can normally solubilize proteins (urea, etc.) and their exact composition is not yet known. However, there is some evidence that they might represent abnormally cross-linked neurofilament proteins. Exactly how these abnormal cytoskeletal structures contribute to the dementing process is not yet known.

A second disease in which cytoskeletal structures appear to participate is liver damage due to alcoholism where aggregates (called Mallory bodies) of short, straight rods, each 14–20 nm in diameter, accumulate in the cytoplasm of damaged liver cells. These rods contain proteins which appear to belong to the keratin family, but again their role in the disease process is not understood.

Antibodies against various intermediate filament proteins can also be used in immunohistochemistry to examine human cancers and to decide which cell types have turned malignant, e.g. glial fibrillary acid protein antibodies can be used to stain brain tumours derived from astrocytes and antibodies against cytokeratins can be used to classify various cancers of the lung and skin.

Tissue-specific enzymes in medicine

One of the earliest applications of enzyme measurement in medicine dates from 1908, when amylase was assayed in urine as an index of the severity of acute pancreatitis (a condition of acute inflammation of the pancreas leading to release of digestive enzymes into the circulation and thus a high mortality rate). During the 1930s and 1940s, measurements of alkaline phosphatase activity in serum became established as a clinically useful test in various bone and liver diseases. A watershed came in 1954, when it was demonstrated that patients with myocardial infarctions (heart attacks) showed marked rises in glutamate-oxaloacetate transaminase (aspartate transaminase) activity in serum following the acute event. This led to the realization that monitoring the activity of intracellular enzymes released into the circulation after tissue damage could indicate the extent of the tissue damage and also which organ was involved. At about the same time, reliable UV spectrophotometers for use in hospital laboratories became available (hence, many enzymes could be coupled to NAD and NADP) and subsequently at least 100 different enzymes have been studied as potential diagnostic markers in an enormous variety of different disease states. The average hospital laboratory will measure at least a dozen different enzymes as diagnostic markers and in addition will use many different enzymes as specific reagents to assay non-enzymic substances in blood (say, 150000 tests a year in this category). World-wide at least 100 million serum measurements of creatine kinase alone are performed each year (mainly to diagnose heart attacks) and approximately 50 million total enzyme measurements are done annually in the UK, i.e. on average, each member of the population has an enzyme measured in their blood each year.

Many intracellular enzymes occur within cells at high concentrations. For instance, several of the 15 enzymes of glycolysis (e.g. aldolase and enolase) are present in tissues at concentrations in excess of $200 \mu g \, g^{-1}$ wet weight and in skeletal muscle the enzyme creatine kinase forms up to

10 % of the soluble cytoplasmic protein. Some enzymes show a characteristically specific tissue distribution and occur at high levels only in particular organs: e.g. acid phosphatase is found at high concentrations in the prostate gland; amylase is found in high concentrations only in the pancreas and salivary glands. The tissue distribution of some clinically important enzymes is shown in Table 3.2.

When a particular tissue is damaged by inflammation, by trauma or by having its blood supply cut off (ischaemic damage), the plasma membrane becomes permeable to intracellular substances which leak out into the general circulation. At first, ions and molecules of low molecular weight (e.g. nucleotides) escape, but if the damage to the cell is severe enough, these are followed by large macromolecules including intracellular enzymes. Although the levels of enzymes found in serum may be 1000-fold less than the levels in the tissues of origin, many enzymes are found in serum in high enough concentrations to be measurable as specific enzyme activities. An example of leakage from a damaged tissue is shown in Fig. 3.8. Here the patient had an attack of infectious hepatitis, a viral disease which leads to acute inflammation of the liver and hence jaundice. The

Table 3.2 *Principal tissue localization of some diagnostically important enzymes*

Enzyme	Main localization
Acid phosphatase	Prostate erythrocytes, lysosomes
Aldolase	Skeletal muscle, heart
Alkaline phosphatase	Bone (osteoblasts), intestinal mucosa, liver, placenta, kidney
Amylase	Pancreas, saliva
Arginase	Liver
Acetylcholinesterase	Brain, nervous tissue, erythrocytes
Alanine aminopeptidase	Kidney, intestine
Alanine transaminase	Liver, skeletal muscle, heart
Aspartate transaminase	Heart, liver, skeletal muscle, kidney, brain
Creatine kinase	Skeletal muscle, heart, brain
Glucose-6-phosphatase	Liver
Isocitrate dehydrogenase	Liver
Glutamate dehydrogenase	Liver
Lactate dehydrogenase	Heart, liver, skeletal muscle, kidney erythrocytes, pancreas, lung
5'-nucleotidase	Hepatobiliary tract, pancreas
Ornithine carbamoyltransferase	Liver
Trypsin(-ogen)	Pancreas

Data from Wilkinson, J. H. (1976). *The Principles and Practice of Diagnostic Enzymology*. Edward Arnold, London.

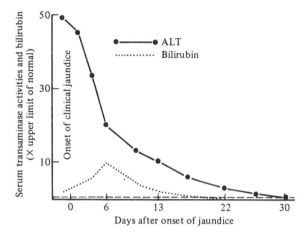

Fig. 3.8. Measurement of a liver enzyme in a patient with jaundice. Alanine (or glutamate-pyruvate) transaminase (ALT) is present at high levels in the cytoplasm of the hepatocyte. Here the enzyme activity was measured serially in the serum of a patient with viral hepatitis. Note the very high level of the enzyme in the bloodstream *before* the appearance of jaundice due to the rising bilirubin level. Data source as for Table 3.2.

enzyme glutamate-pyruvate transaminase (alanine transaminase) occurs at high concentrations in the cytoplasm of liver cells. As the inflammation proceeds, large quantities (up to half of the total liver content of the enzyme) are released into the circulation, producing astronomical levels of alanine transaminase activity in the patient's serum. As the patient gets better and the inflammation resolves, the release of the enzyme ceases and the serum level transaminase activity falls back to normal values (Fig. 3.8). Note that very high levels of transaminase activity can be found in the patient's serum before the onset of jaundice, i.e. by measuring enzyme activity, the diagnosis can be made very early in the disease.

Isoenzymes

As more and more enzyme activities were studied and catalogued, it became obvious that many enzymes existed in more than one form, i.e. different protein molecules exhibited the same enzymic activity. In 1959, Market and Mollert introduced the term 'isozyme' or 'isoenzyme' for sets of proteins with the same enzyme activity. The early detection of isoenzymes was dependent on electrophoretic separation, followed by specific staining procedures for the enzyme activity. Two or more bands of enzyme activity showed the presence of isoenzymes (see example in Fig. 3.9). Since this separation depends on charge differences,

(*a*) Alkaline phosphatase

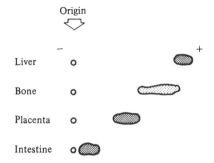

Electophoretic separation of human tissue alkaline phosphatases on agar-gel.

(*b*) Lactate dehydrogenase

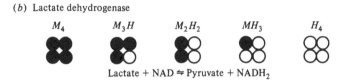

Lactate + NAD $\rightleftharpoons$ Pyruvate + NADH$_2$

(*c*) Creatine kinase

Creatine + ATP $\rightleftharpoons$ Creatine $\sim$ P + ADP

Fig. 3.9. Some isoenzyme systems. The members of each isoenzyme set are shown with the more acidic isoenzymes to the right. The major *in vivo* substrate of alkaline phosphatase is unknown. The enzyme will remove phosphate groups from a wide variety of biological substrates. Data source for (*a*) as for Table 3.2.

isoenzymes with different primary sequences but with the same charge are not detected and early studies underestimated the incidence of isoenzymes in biological tissues. It is now clear that probably the majority of enzymes exist in multiple forms, e.g. out of the 10 enzymes involved in the conversion of glucose to pyruvate, 9 are known to exist as isoenzymes and in particular as different isoenzymes in liver and muscle.

The majority of isoenzymes can be classified into three structural classes. The first category is multi-subunited isoenzymes whose individual subunits are coded for by separate genetic loci. An example is lactate dehydrogenase (LDH), which is a tetrameric (four-subunited) enzyme composed of a mixture of muscle-type (*M*) and heart-type (*H*) subunits. In adult skeletal muscle, the major form of LDH is an M_4

tetramer, in adult heart an H_4 tetramer (see Fig. 3.9(b)). Other tissues show a mixture of these two 'parental' forms of the enzyme, hence five LDH isoenzymes are possible: M_4, M_3H, M_2H_2, MH_3 and H_4 (Fig. 3.9(b)). Aldolase is another tetrameric isoenzyme 'set', with A subunits characteristic of adult skeletal muscle, B subunits characteristic of adult liver, and C subunits which are largely confined to adult brain tissue. Hence, in liver, aldolase is composed of four B subunits, the B_4 isoenzyme, skeletal muscle contains the A_4 isoenzyme, and in brain (because the A subunit is also present), aldolase is a 'hybrid set' of A_4, A_3C, A_2C_2, AC_3 and C_4 isoenzymes. Although the primary amino acid sequence of different parental subunits of multi-subunited isoenzymes can show a high degree of homology (indicating that they have evolved by gene duplication), antisera (especially polyclonal antisera) can readily distinguish between them. For example, an antiserum raised against aldolase A will not react with aldolase B_4, and an antiserum raised against LDH M_4 will not react against LDH H_4.

The second major structural group of isoenzymes consists of those with a single subunit (monomeric isoenzymes), again coded for by separate genes and having different primary amino acid sequences. An example is aspartate (glutamate-oxaloacetate) transaminase, which exists as two separate monomeric isoenzymes, one located in the cytoplasm and one in the mitochondrion. The third structural class of isoenzymes is of those which show post-translational modifications of the same primary sequence, as is probably the case with alkaline phosphatase isoenzymes.

Many isoenzyme sets show characteristic 'switches' from one type of isoenzyme to another during development. For example, skeletal muscle shows heart-type LDH H_4 and liver type enolase in embryonic life, but these are replaced by LDH M_4 and muscle-type enolase by the time adulthood is reached. The reason for this metabolic 'switching' is usually obscure. Several cancers are known to produce atypical (sometimes foetal-type) isoenzymes, reflecting their de-differentiated state.

Why do isoenzymes exist?

While the existence of multiple forms of enzymes in different tissues is well established, convincing explanations of the biological significance of this phenomenon and of the evolutionary pressure which has brought it about are often lacking. Isoenzymes sometimes have different kinetic properties which can be clearly related to their biological role. An example of this is seen by comparison of hexokinase – found in liver and muscle, *Km* for glucose 0.1 mM, inhibited by the product of the reaction, glucose 6-phosphate – and glucokinase – found only in liver, *Km* 10 mM, not inhibited by glucose-6-phosphate. One of the major functions of the liver is to 'mop-up' plasma glucose and store it as glycogen, hence glucokinase can process glucose presented to it by the bloodstream (normal concentration 5 mM) without being inhibited by the accumulation of glucose-6-phosphate. In muscle, however, the entry of glucose is controlled by the presence of insulin and the intracellular concentration of glucose is 0.1 mM. Therefore the presence of the glucokinase isoenzyme in liver can be related to its metabolic role. A second possible example of different kinetic properties lending credence to the existence of isoenzymes is seen with aldolase isoenzymes. The liver-type aldolase B_4 isoenzyme has a V_{max} for FBP synthesis 10 times higher than the V_{max} for FBP cleavage, while for the muscle-type aldolase A_4 isoenzyme the maximum velocity is only twice as high. This has been proposed to reflect the gluconeogenic role of aldolase in liver which is absent in muscle.

Such 'kinetic' explanations for the existence of isoenzymes (together with those based on 'compartmentation' or on differential sensitivity to inhibitors, etc.) can be marshalled to explain only a minority of the defined cases in which isoenzyme sets are known to exist. There remain a large number of enzymes which are distinguishable on the basis of charge, immunological properties and primary sequence, but which appear kinetically identical. Several speculative proposals for the existence of this unexpectedly high occurrence of multiple forms of enzymes have been made. One suggestion is that metabolic pathways in the 'soluble' cytoplasmic compartment of cells may not occur in free solution but may take place on organized multimolecular assemblies of enzymes, each representing a particular and separate metabolic route. This is illustrated in Fig. 3.10. Such 'polyisozymic complexes' need multiple forms of enzymes to enable this structural assembly of different combinations of enzymes to take place. Another suggestion is that different forms of enzymes are needed in different cells to interact with different cytoskeletal elements.

Clinically important isoenzymes

Whatever the biological explanation for the widespread occurrence of multiple molecular forms of enzymes, such isoenzymes have proved clinically very useful as diagnostic markers. In adult human tissues, specific distributions of isoenzymes enable accurate assessment of the site of tissue damage by measuring specific isoenzymes in the circulation. Since individual members of many clinically useful isoenzyme sets are kinetically indistinguishable, electrophoretic or immunological methods are often used on serum samples to detect particular isoenzymes. We

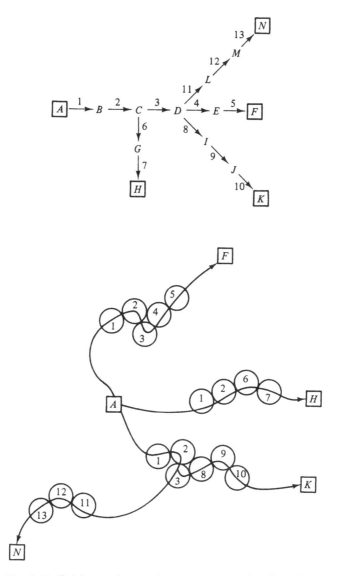

Fig. 3.10. Polyisozymic complexes as a reason for the existence of isoenzymes. In this example, a hypothetical substrate *A* can be converted to *F*, *H*, *K* or *N*, depending on the metabolic pathway taken, each step being catalysed by an enzyme (1 to 13). It is postulated that each part of the pathway is on a separate multienzyme complex as shown. In cases where the same enzyme is present in different complexes (e.g. enzymes 1, 2 and 3 are in complexes 1–5, 1+2+3+8+9+10 and 1+2+3+11+12+13), it is proposed that different isoenzymes, e.g. 1*a*, 1*b* and 1*c*, would be present in different complexes.

shall select three isoenzyme sets which are widely used as diagnostic markers in clinical situations: creatine kinase, lactate dehydrogenase and alkaline phosphatase.

The creatine kinase (CK) set of cytoplasmic isoenzymes consists of the three dimers, CK–MM, CK–MB and CK–BB. (There is also a mitochondrial form of creatine kinase coded for by a third gene.) CK–MM is characteristic of adult human skeletal muscle, CK–BB is characteristic of brain, while the hybrid dimer CK–MB is virtually confined to cardiac muscle where it forms about 15 % of the total creatine kinase (the rest being CK–MM). In normal human serum, detectable creatine kinase activity is largely due to CK–MM, with only very low levels ($1 \mu g/l$) of CK–MB and CK–BB present. Following a myocardial infarction (heart attack), CK–MB is released from damaged cardiac muscle and this isoenzyme appears in serum along with increased amounts of CK–MM (see Fig. 3.11). Other enzymes are also released from cardiac muscle. These include aspartate (glutamate-oxaloacetate) transaminase (AST) and 'heart-type' H_4 lactate dehydrogenase 3.11 (LDH) (see Fig. 3.11). This particular isoenzyme of LDH (also known as LDI) shows greater activity towards β-OH-butyrate than towards lactate and is therefore measured as 'β-OH-butyrate dehydrogenase' activity. Although creatine kinase, aspartate transaminase and LDH H_4 all originate from damaged cardiac muscle following a myocardial infarction, the appearance of each

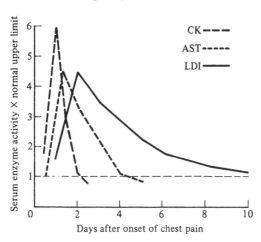

Fig. 3.11. The pattern of enzyme release from heart muscle following a myocardial infarction. See text for details. LDI is an alternative name for the heart-specific H_4 LDH isoenzyme (see Fig. 3.9). This isoenzyme preferentially uses β-hydroxybutyrate rather than lactate and is often referred to as β-hydroxybutyrate dehydrogenase. Data source as for Table 3.2.

enzyme activity in serum follows a characteristic temporal pattern as shown in Fig. 3.11. Creatine kinase activity reaches a maximum rapidly (within 24 hours), followed by aspartate transaminase activity (1–2 days), followed by LDH activity (3–4 days). This pattern of activity arises as a result of differential rates of release from cardiac muscle and differential rates of clearance of each enzyme from the circulation, neither of which are clearly understood. However, in certain circumstances, the characteristic pattern of serum enzyme changes shown in Fig. 3.11 can be a more reliable indication of myocardial infarction than can the electrocardiogram.

Alkaline phosphatase belongs to the class of enzymes known as 'ectoenzymes' and its true physiological role in the body is unknown. However, alkaline phosphatase is a diagnostically important enzyme which has been measured in clinical laboratories for at least 50 years. There are four major isoenzymes of alkaline phosphatase, derived from liver, bone, placental and intestinal tissue, respectively (see Fig. 3.9). It is currently not clear how different these are in primary sequence or whether their major differences are due to post-translational modification. The most important use of liver alkaline phosphatase is in the diagnosis of liver disease, especially in cases of jaundice due to obstruction of the biliary outflow tract, e.g. by gall stones. Bone-type alkaline phosphatase is measured in a variety of skeletal disorders, e.g. raised levels are seen in the serum with bone diseases such as rickets. Occasionally, an abnormal form of alkaline phosphatase isoenzyme (the Regan isoenzyme) is found in the serum of patients with certain cancers, especially with carcinoma of the bronchi. This isoenzyme resembles the placental-type isoenzyme and is an example of abnormal isoenzyme production by a malignant tumour.

Enzymes as diagnostic reagents

Enzyme specificity for measuring drugs in plasma
An example is an assay for paracetamol, an analgesic often taken by the 20 000 patients admitted to hospital in the UK each year with drug overdoses. If treated within 10–12 hours of the overdose, the severely poisoned patient can recover; if not, irreversible liver damage occurs. The treatment itself (cystaemine) is toxic and causes heart arrhythmias, therefore it is necessary to know the level of paracetamol in the circulation to know whether treatment is justified or not. To measure this, a bacterial enzyme is used which hydrolyses paracetamol (*N*-acetyl *p*-amino-phenol) to acetate and *p*-aminophenol, a reaction which forms a coloured product with a dye.

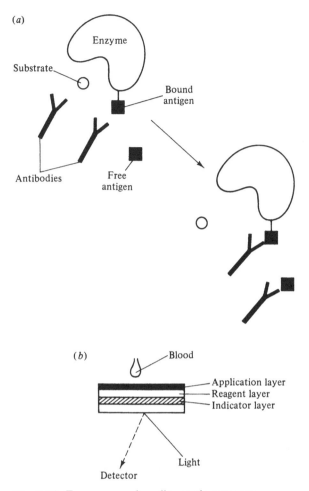

Fig. 3.12. Enzymes used as diagnostic reagents.

(a) Enzymes as labels in immunoassay systems. A scheme is shown where a specific antigen is coupled to an enzyme. When a specific antibody to the antigen is added, the binding of the antibody to the antigen inhibits the catalytic activity of the enzyme (possibly by steric hindrance preventing the substrate reaching the active site). The system can be used to measure the amount of unknown antigen in a sample, since the more of this present, the less antibody is available to inhibit activity by binding to the antigen coupled to the enzyme. Alternatively, enzymes can be coupled directly to antibodies (cf. immunohistochemistry p. 55), when the more enzyme activity measured, the more antigen is present binding the coupled antibody.

(b) Thin-film technology. A spot of blood sample (e.g. from a diabetic wishing to check his blood sugar) is applied to the top of the film. Glucose in the sample diffuses through to the reagent layer (e.g. containing the enzyme glucose oxidase), and a product results which diffuses through to the indicator layer. Changes in the indicator can be seen by a reflectance meter.

Enzymes as labels in immunoassay systems
These types of assays, illustrated in Fig. 3.12, are named enzyme-linked immunoadsorbent assay (ELISA) and enzyme-multiplied immunoassay techniques (EMIT). They are especially useful for measuring drugs and large antigens and in situations where radioactive techniques are unavailable (e.g. in Third World countries).

Immobilized enzymes, enzyme electrodes and thin-film technology
Enzymes can be immobilized on to solid supports with retention of enzyme activity and specificity. This means that the enzyme can then be used in continuous-flow analysis machines such as the prototype artificial pancreas. Enzyme electrodes can be designed in which enzyme activity is monitored as an electrical current, e.g. the glucose electrode which produces a current directly proportional to the glucose concentration. Thin-film reagent strips can be made (Fig. 3.12) with which concentrations of specific substances in blood can be monitored with small reflectance meters, e.g. the glucometer for home use by diabetics. These newer enzyme techniques mean that biochemical measurements of clinical importance are moving out of the laboratory and closer to the patient.

Enzymes as therapeutic agents
Enzymes in cancer therapy
In the early 1950s, experiments were being carried out attempting to treat lymphoma (a tumour of the lymphatic system) in mice by injecting antibodies against the tumour tissue into the whole animal. Along with the anti-tumour antibodies, normal guinea-pig serum was also injected as an intended source of complement. Surprisingly, mice receiving guinea-pig serum alone also showed regression of the tumour. Guinea-pig serum is unusual among mammalian sera in containing asparaginase, an enzyme which catalyses the following reaction:

$$L\text{-asparagine} + H_2O \rightarrow L\text{-aspartate} + NH_3$$

Some tumour cells require the non-essential amino acid asparagine as a growth factor because, unlike normal cells, they contain insufficient asparagine synthetase activity to satisfy their metabolic needs. The effect of the normal guinea-pig serum was to deplete circulating levels of asparagine, so that the tumour cells died while normal cells survived.

Asparaginase is now used, usually in combination with chemotherapeutic agents, to produce remission in acute lymphoblastic leukaemia, a form of the disease seen in children. Other non-essential amino acids, of which the most promising are glutamine, arginine, cysteine and citrulline, are also candidates as 'growth factors' for tumour cells and possible therapeutic actions of enzymes which can degrade these in the circulation are being actively tested.

While the therapeutic action of asparaginase is an important practical demonstration of an enzyme-induced nutritional deficit specific for tumour cells, there are several practical problems in the intravenous use of enzymes in this way. These include rapid clearance of the enzyme from the circulation (which can sometimes be slowed by chemically modifying the enzyme to alter its net charge or by polymerizing the enzyme to increase its molecular weight) and the development of antibodies against the enzyme, since most enzymes (including asparaginase) used in this way are derived from bacterial sources. A possible way of avoiding both of these problems may be to immobilize the enzyme on a solid support contained in some form of extracorporeal circulation device.

Enzymes as thrombolytic agents

The clotting process is a complicated cascade of proteolytic events which eventually leads to the cleavage of fibrinogen to fibrin and the cross-linking of this into an insoluble matrix which forms the basis of the blood clot or thrombus (Fig. 3.13). Thrombosis within an artery can obliterate the blood supply distal to the obstruction with catastrophic effects, e.g. myocardial infarction (coronary artery thrombosis), a stroke (cerebral arterial thrombosis), or the occlusion of an artery to a limb leading eventually to gangrene. It is also possible for a thrombus to form in one part of the circulation (e.g. the chambers of the heart) and then detach and lodge elsewhere in the circulation (an embolus). This is particularly common in the venous side of the circulation, where a dislodged thrombus passes to the lungs (a pulmonary embolus). Thromboembolic vascular disease is the commonest single cause of death in later years in Western populations. Early dissolution of a thrombus or embolus can restore circulation and prevent tissue necrosis.

Plasminogen is a serum protein which acts as a precursor for plasmin, a proteolytic enzyme which hydrolyses proteins at lysyl and arginyl peptide bonds. Plasminogen, when activated, undergoes a specific cleavage at a single Arg–Val bond in the interior of the molecule and a second cleavage near the N-terminus releasing a small peptide. This converts plasminogen to the active protease plasmin, which is made up of two polypeptide chains (of molecular weights 48 kDa and 27.5 kDa) linked by a single disulphide bond. Plasmin is not specific for fibrin, but will hydrolyse a variety of proteins and synthetic peptidyl substrates. The protease activity of any free plasmin in plasma or serum is neutralized by the serum

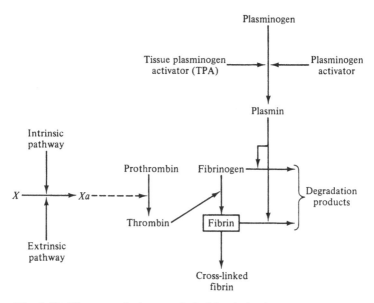

Fig. 3.13. The proteolytic cascade in blood clotting. The conversion of fibrinogen to fibrin needs the proteolytic enzyme thrombin, which is formed from prothrombin by activated factor X (Xa). The conversion of factor X to its activated form (Xa) can occur by the 'intrinsic' or the 'extrinsic' pathway. Both pathways are a complicated cascade of proteolytic events, the intrinsic pathway being activated by sub-endothelial contact at the site of blood vessel injury, the extrinsic pathway by factors released on tissue damage. Fibrinogen and fibrin can be removed (and blood clotting prevented) by plasmin. This proteolytic enzyme is formed from plasminogen by tissue plasminogen activator (TPA) or by plasminogen activator present in plasma.

anti-protease system including α_1-antitrypsin and α_2-macroglobulin (see p. 23). There are various activators of plasminogen which can initiate the conversion to active plasmin. Most tissues contain tissue plasminogen activator (TPA), which is probably identical to the plasminogen activator found in normal plasma. Urokinase (MW 54 kDa) is a plasminogen activator found in normal urine and is probably distinct from tissue plasminogen activator. Urokinase is a protease which can directly cleave plasminogen. Streptokinase (MW 47 kDa) is derived from the haemolytic streptococcus and does not cleave plasminogen directly but forms a complex with it. The formation of this complex results in plasmin being released and the streptokinase itself being fragmented. Streptokinase will activate plasminogen from humans, cats, dogs and rabbits but not from horses, sheep, cows and rats.

On the formation of a fibrin clot in the circulation, a small proportion of total circulating plasminogen is trapped within the fibrin matrix, while most plasminogen remains free in the circulation. Both urokinase and TPA appear to have a much greater affinity for plasminogen absorbed in a fibrin clot than for plasminogen free in the circulation. Therefore plasmin is released within the clot and can act on fibrin in an environment relatively free of the usual circulating protease inhibitors. Streptokinase, however, does not have a preferential affinity for clot plasminogen and reduces total circulating plasminogen to low levels before therapeutic clot lysis occurs. Additionally, since most of the population have been exposed to the streptococcus, antibodies against streptokinase are usually already present in the patient and administration of streptokinase produces a rapid rise in antibody titre and neutralization of the therapeutic effect. Urokinase is a human protein and is therefore not antigenic, but it has the disadvantage that it is more rapidly cleared from the circulation.

Streptokinase can be obtained in a pure form in large quantities from bacterial sources and is readily available. Urokinase can be purified from human urine but large volumes are needed. Recently, tissue cultures of human embryonic kidney cells have been used as a source. Development of expression vector systems (see Chapter 7) is solving the shortage of these plasminogen activators by genetic engineering methods.

Enzymes as specific detoxifying agents

The high affinity and specificity of catalysis shown by enzymes under physiological conditions should make them suitable for use in cases of poisoning by drugs, toxins, snake venoms, etc. In fact, there are very few examples of enzymes being used in this way. The best is methotrexate (MTX) 'rescue' in the treatment of cancer. Many folate antagonists are available which cause the regression of tumours by inhibiting dihydrofolate reductase. This causes depletion of folate coenzymes and a decrease in the rate of synthesis of thymidine, purine bases and certain amino acids, all of which are necessary for rapid tumour growth. Inducing folate deficiency in patients with cancer can be done by dietary means (although this takes a long time) or more rapidly by giving folate antagonists such as MTX (see the HAT medium trick p. 41). MTX, however, produces toxic side reactions on rapidly dividing cells in the bone-marrow and cells lining the gastrointestinal tract, which appear to be due to the persistence of low levels of MTX within the circulation for long periods of time after the initial administration of the (large) therapeutic dose. Carboxypeptidase G_1 is an enzyme from a *Pseudomonas* bacterium which can cleave the terminal glutamate not only from ^5N-methyltetrahydrofolic acid but also from MTX. This leaves an MTX derivative which is 100 times less effective in inhibiting dihydrofolate reductase. Patients given large doses of MTX to kill their tumour cells can therefore be 'rescued' from the toxic effects of persistent MTX by the administration of carboxypeptidase G_1. Other examples of the use of enzymes as detoxifying agents are the use of uricase to reduce high plasma uric acid levels in patients with leukaemia, and in attempts to develop enzyme-based 'artificial kidneys' to treat patients in renal failure.

Enzyme replacement in inborn errors of metabolism
There are approximately 200 diseases known which are 'inborn errors of metabolism', i.e. diseases which can be explained as a single defect in a specific enzyme in a metabolic pathway. These disorders, which are inherited as autosomal recessive disorders, often involve defects in enzymes normally localized in lysosomes. These enzymes are involved in pathways breaking down complex lipids and polysaccharides. In the presence of defective enzyme, the relevant substrate accumulates, hence the disorders are often described as lysosomal 'storage' disorders. Examples include the lipidoses (Fabry's disease, Gaucher's disease, metachromatic leucodystrophy, Niemann–Pick disease) and the mucopolysaccharidoses (Types I to VII).

The rationale for treating these diseases (which are all relatively rare) has been to replace the defective enzyme either by supplying enzyme protein or by replacing the defective gene with a normal gene. The problems of gene replacement therapy are discussed in Chapter 7. Several attempts have been made to ameliorate various lysosomal storage diseases by supplying the relevant purified enzyme. The purified enzyme is usually made from human placenta, thus avoiding the production of antibodies by the patient against enzyme from non-human sources. Intravenous injection of enzyme results in rapid clearance from the bloodstream, i.e. within minutes or a few hours. In addition, enzymes injected into the general circulation do not appear to cross the blood–brain barrier and some lysosomal disorders, e.g. Tay–Sachs and Niemann–Pick diseases, have severe neurological involvement. Entrapment of enzymes in liposomes or erythrocyte ghosts has been used to slow the clearance of the enzyme from the circulation. This still leaves the problem of being able to target the entrapped enzyme to the tissue where the defect needs to be corrected, e.g. liver, spleen, muscle or bone marrow. Attempts are being made to promote specific receptor-mediated endocytosis by attachment of enzymes, e.g. to low-density lipoproteins

(LDL), and by coating red cells with appropriate carbohydrate recognition signals. Much further research on tissue-specific uptake mechanisms will be necessary before clinically effective enzyme replacement therapy by these sorts of methods is a reality.

4

The plasma membrane

The plasma membrane as the link between the cell and its environment

The surface of all living cells is a membrane, the plasma membrane, which acts as a boundary between the cytoplasm and the environment, limiting the diffusion of substances into and out of the cell and protecting the cell from adverse or fluctuating environmental conditions. It is responsible for such diverse functions as control of cellular nutrition, cell–cell communication, cell adhesion and immunogenicity. Its properties are largely defined by a common structure consisting of a hydrophobic lipid core, unable to allow the passage of water and charged molecules, punctuated by protein pores, receptor proteins and enzymic proteins which confer specificity to the transport and regulatory functions of the membrane. The growth of our knowledge of its structure and dynamics has gradually allowed a recognition of its fundamental importance in mammalian cells to such activities as hormone and neurotransmitter action, drug action, cell damage in disease processes and cell death.

The plasma membrane as an osmotic barrier surrounding cells

The origin of the concept and term plasma membrane is intimately linked to the development of cell theory and the use of light microscopy as an experimental tool in the nineteenth century. Over the same period, as it was being recognized that all living tissues were composed of cells and that new cells could only be formed by division of pre-existing cells, it was also slowly acknowledged that there existed a surface boundary layer endowing the cell with its osmotic properties. This was first apparent in the studies of a number of scientists in Germany, who were able to show the osmotic properties of plant cells (Fig. 4.1). From such studies, Nageli called the cell surface layer the *Plasmamembran*, defining it as a surface layer of protoplasm, denser and more viscous than the protoplasm as a whole. Pfeffer argued that it had properties resembling those of artificial,

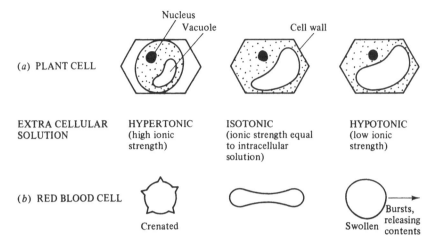

Fig. 4.1. The effect of the osmolarity of an extracellular solution on the structure of a plant cell (*a*) and human red blood cell (*b*). Outside the plasma membrane, the plant cell is surrounded by a cellulose cell wall, which limits its expansion in hypotonic solution and from which it retracts in hypertonic solution. The human red blood cell has no nucleus, is biconcave in cross-section and is not protected by a cell wall. Thus, it swells and bursts in hypotonic solution and becomes crenated in hypertonic solution.

chemical, semi-permeable osmotic membranes. Similar studies showing the osmotic properties of red blood cells (Fig. 4.1) extended the concept of the plasma membrane to mammalian cells, although since mammalian red blood cells act as irregular osmometers, there was much dispute as to the exact significance of these observations. The argument for a plasma membrane in animal cells was strengthened by the work of Chambers in the 1920s, who developed microdissection techniques to puncture the cell surface of amoebae and sea urchin eggs. He also showed that it was possible to remove the cell surface of these cells by microdissection and, using a light microscope, to watch it re-form. However, it was not until the first attempts at defining the thickness and composition of the plasma membrane using analytical chemical and physical techniques that an understanding of the basis of its universal structure and function was possible.

The lipid bilayer as the basic structure of the plasma membrane
It is now known that all biological membranes contain lipid and protein constituents, the major lipids being phospholipids. In 1925, Gorter and Grendel performed a classic experiment which suggested that the basic

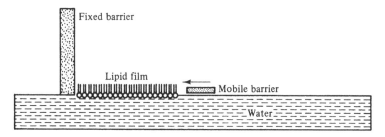

Fig. 4.2. The Langmuir trough used for compression of a lipid film.

structure of the plasma membrane was a bilayer of phospholipid. Mature mammalian red blood cells have no intracellular membrane systems and no nucleus. Gorter and Grendel extracted the lipid from a defined number of red blood cells in acetone, dissolved the evaporated residue in benzene and spread it as a film on the surface of water in an apparatus known as a Langmuir trough (Fig. 4.2). The area of this film of lipid was slowly reduced by sliding a glass barrier across the surface, forcing the lipid against an opposite retaining barrier, and as soon as the film started to exert a measurable resistance to compression, the area occupied by it was measured. This area corresponded to a coherent monomolecular film or monolayer and was found to be twice the surface area of the red blood

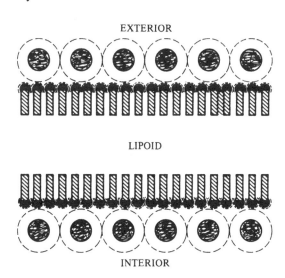

Fig. 4.3. The Davson–Danielli model of membrane structure. Proteins are shown as hydrated globular molecules on each surface of a phospholipid bilayer. Reprinted from Danielli, J. F. & Davson, H. (1935). *J. Cell. Comp. Physiol.*, **5**, 495–508.

cells from which the lipid was derived. The experiments have since been criticized because Gorter and Grendel overestimated the acetone extraction of lipid and underestimated the cell surface area by calculating it from the dimensions of dried red blood cells. Fortunately, these errors cancelled out and the correct conclusion that the lipid constituents of the plasma membrane are arranged as a bimolecular leaflet was included as a central feature of the Davson–Danielli model of membrane structure (Fig. 4.3) which was proposed in the 1930s and dominated the biochemical literature for more than a quarter of a century. This model also suggested the organization of membrane protein and, largely because of the known surface tension and electrical properties of membranes, indicated that it was distributed on either side of the bilayer, associated with the charged head groups of the phospholipids.

The lipid components of the plasma membrane

Membrane lipids are almost all amphipathic, containing a hydrophobic hydrocarbon region and a smaller hydrophilic region called the polar head group. The major lipids in membranes are phospholipids and the sterol lipid cholesterol, the general structures of which are shown in Fig. 4.4. The hydrophobic moiety of phospholipids is derived from long-chain fatty acid molecules which are esterified via a glycerol molecule and a phosphate ester to a hydrophilic base. Most of the fatty acid molecules are simple, unbranched hydrocarbon chains 14–24 carbons long, which may be saturated or unsaturated. In mammalian membranes, one of the two fatty acid substituents of phospholipids is usually saturated and one unsaturated, containing one or more *cis*-double bonds. The degree of

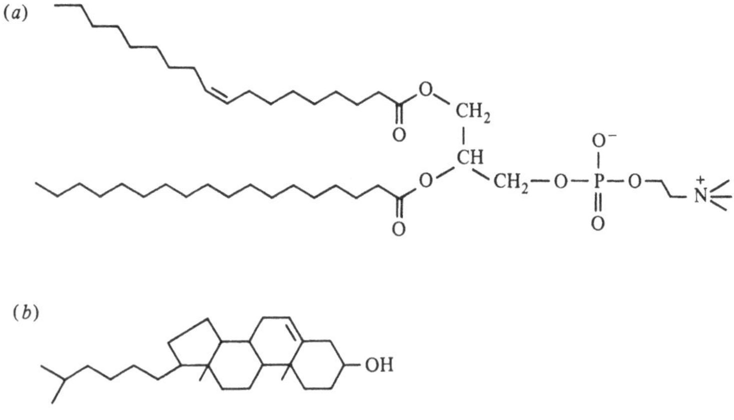

Fig. 4.4. The molecular structure of a phospholipid (*a*) and a sterol (*b*).

unsaturation can have an important effect on the packing and fluidity of phospholipid molecules in the membrane.

The different classes of phospholipid defined by their different hydrophilic head groups can be separated and analysed by thin-layer chromatography (TLC). Conventionally, chromatography separates different chemical substances on the basis of their partition between a moving solvent phase and a stationary phase (e.g. water in the paper during paper chromatography). TLC is an experimentally easy and rapid technique particularly suited to separating membrane lipids on a thin layer of silica or alumina formed on a glass plate. The chromatogram is run (developed) in a solvent mixture and the phospholipids separated by partition between this phase and the stationary silicic acid. Phospholipids are visualized after drying the plate by exposure to an atmosphere of iodine, which reacts with the unsaturated fatty acid substituents producing yellow spots. The spots are identified by comparison with standards and may be quantified by extraction of the lipid and measurement of phosphorus content. A typical separation showing the major phospholipids of a rat liver cell plasma membrane is shown in Fig. 4.5, together with the chemical structure of their head groups. Further analysis of the fatty acid chains of the separated phospholipids can be achieved using more complex techniques such as gas–liquid chromatography (GLC), where individual volatilized, derivatized fatty acids partition between a moving gaseous phase (usually argon or nitrogen) and a liquid phase coated on a finely divided solid support (Fig. 4.6). The system is usually run at temperatures of 100–250 °C when the various greases and esters used as the liquid phase melt.

Cholesterol is a major constituent of the mammalian plasma membrane constituting approximately 45 % of the total lipid. The hydroxyl group provides the polar head group and the four ring structure the hydrophobic tail (Fig. 4.4). Although not included in the earliest models of membrane structure, it is now thought that it plays a vital role in stiffening the lipid bilayer, the hydrophobic steroid ring interacting with the hydrophobic fatty acid chains of the phospholipids (Fig. 4.7).

Other lipids in addition to phospholipids and cholesterol are also present in the plasma membrane as minor components and differ between tissues and cell types. These include plasmalogens, gangliosides and glycosphingolipids, which have structural similarities to phospholipids. The gangliosides in particular are of functional importance since they can act as receptors for some hormones and toxins and define immunogenicity on specific cell types.

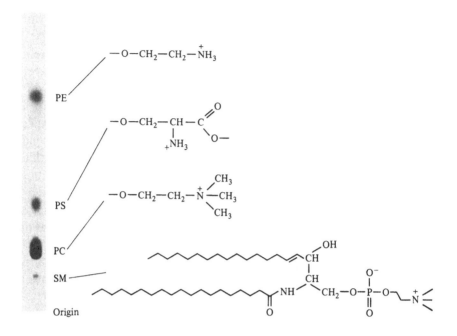

Fig. 4.5. Separation of the major phospholipids of rat liver plasma membrane by thin-layer chromatography on silica gel plates. The solvent system used was chloroform/methanol/acetic acid/water (45:25:10:1 by volume). The molecular structures of the different polar head groups of the phospholipids, phosphatidyl ethanolamine (PE), phosphatidyl serine (PS) and phosphatidyl choline (PC) are shown. The molecular structure of sphingomyelin (SM) differs from the basic phospholipid structure shown in Fig. 4.4 and is shown in full.

Lipid asymmetry in the plasma membrane

It is possible to show that the various species of phospholipid are not evenly distributed between the two halves of the bilayer in the plasma membrane. This asymmetric distribution was first shown in the case of the red blood cell membrane by reaction with non-penetrating reagents capable of interacting with specific phospholipids. It was found that phosphatidyl ethanolamine (PE) reacted far more readily with formylmethionylsulphone methyl phosphate (FMMP, which reacts with aminophospholipids) in red blood cell ghosts, produced by hypotonic lysis of cells, than in intact cells. It therefore appeared that PE was preferentially localized in the cytoplasmic half of the bilayer. Phospholipases, enzymes capable of degrading specific phospholipids, are also

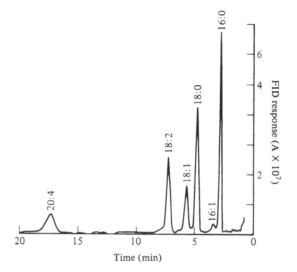

Fig. 4.6. Gas–liquid chromatography separation of fatty acid chains from rat liver plasma membrane phospholipids. Separation was achieved on a column of ethylene glycol succinate at 180 °C. FID, flame ionization detector. Reprinted from Stanley, K. K. & Luzio, J. P. (1978). *Biochim. Biophys. Acta*, **514**, 198–205.

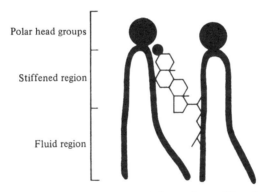

Fig. 4.7. The function of cholesterol to stiffen the lipid bilayer. A representation of half the bilayer is shown.

non-penetrating reagents due to their large protein molecular weight and can therefore be used to determine the relative amounts of phospholipids in the two halves of the bilayer by comparison of phospholipid hydrolysis in intact cells and leaky ghosts. Thus, it is now known that sphingomyelin (SM) and phosphatidyl choline (lecithin, PC), the major phospholipids (>50 %) of mammalian red blood cell membranes are mostly in the outer

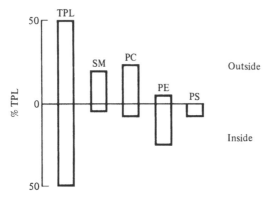

Fig. 4.8. Phospholipid asymmetry in the human red blood cell membrane. TPL, total phospholipid; SM, sphingomyelin; PC, phosphatidyl choline; PE, phosphatidylethanolamine; PS, phosphatidylserine.

half of the bilayer while phosphatidyl serine (PS) like PE is in the inner half of the bilayer (Fig. 4.8). Cholesterol is thought to be in both halves of the bilayer, possibly with a greater proportion associated with the outer half of the bilayer. Similar experiments have been carried out with intact and leaky membrane preparations from other cells for both plasma membrane and intracellular membranes and, in general, have shown asymmetric distribution of the major phospholipid species.

The protein components of the plasma membrane
The relative amounts of protein and lipid in the plasma membrane vary between cell types, the protein content being 40 % by weight in red blood cells but only 25 % by weight in the myelin sheath. In general, membrane proteins can be classified as peripheral (extrinsic) membrane proteins and integral (intrinsic) membrane proteins. The former can be isolated from the membrane by treatment with high salt concentration, whereas the latter contain hydrophobic portions embedded in or transversing the bilayer and can only be solubilized by treatment with detergents. The recognition of the existence of integral membrane proteins meant that the Davson–Danielli model of plasma membrane structure required modification so that the bilayer is disrupted by transmembrane proteins (Fig. 4.9). Clearly, integral proteins may in principle be transmembrane, or associated with only the inner or outer half of the bilayer. There are now many known examples of transmembrane proteins, including glycophorin and Band III protein in red blood cells, Na^+, K^+ ATPase and the insulin receptor. Some examples are known of proteins associated

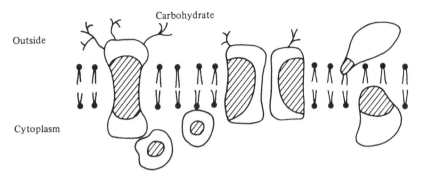

Fig. 4.9. The association of plasma membrane proteins with the lipid bilayer. Various known interactions are shown diagrammatically. Hydrophobic regions of membrane proteins are hatched.

with the inner half of the bilayer, e.g. adenylate cyclase, and also integral proteins in the outer half of the bilayer. The latter, e.g. the antigen Thy-I in thymocytes, have a covalently linked lipid tail penetrating the bilayer.

Just as the phospholipid content of the plasma membrane can be analysed easily using TLC, it is possible to analyse the protein composition using sodium dodecyl sulphate–polyacrylamide gel electrophoresis (SDS–PAGE). In this procedure, described in Chapter 1, the powerful anionic detergent SDS is used to solubilize the membrane proteins and dissociate protein subunits. The ability of SDS–PAGE to separate membrane proteins is shown in Fig. 4.10. Although SDS treatment is a good means of solubilizing membrane proteins, it usually results in the inactivation of their biological activities. In analytical systems, the SDS can be electroeluted away from the protein with some success, allowing recovery of native properties. Thus, immunoblotting (Western blotting) is often a useful way of further analysing membrane protein mixtures if specific antibodies are available (see Chapter 1).

Despite the enormous advantages of using SDS to solubilize membrane proteins for analytical experiments, most attempts to purify integral membrane proteins have relied on milder detergents of which there are a great variety. Essentially, the aim is to displace membrane lipids from the hydrophobic areas of membrane proteins by the amphipathic detergent molecules to achieve optimal solubilization, defined as producing the smallest and least polydisperse biologically active protein. The difficulties of obtaining the best conditions for this are great, since many stages of solubilization can occur (Fig. 4.11). However, a wide variety of membrane proteins have now been purified by solubilization in mild detergents with subsequent application of the

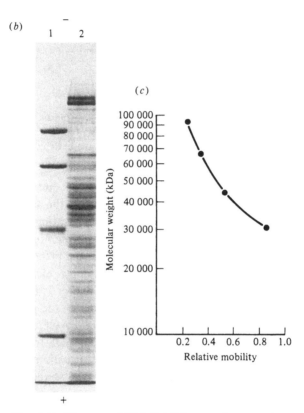

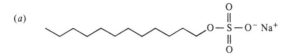

Fig. 4.10. The use of SDS–PAGE to analyse membrane proteins. (*a*) the molecular structure of SDS; (*b*) SDS–PAGE showing, in track 1, the separation of molecular weight markers (from the top of the gel phosphorylase *b*, 92 500, serum albumin, 66 200, ovalbumin, 45 000, carbonic anhydrase, 31 000) and, in track 2, the complex protein pattern obtained from a membrane pellet prepared by centrifuging a liver homogenate at 10 000 g for 10 min. Proteins were visualized by staining with Coomassie blue which allows major proteins to be seen. The molecular weights of these proteins can be estimated by comparison with the markers shown in track 1, whose migration may be plotted as a standard curve (*c*). Data in (*b*) reprinted from Stanley, K. K. & Luzio, J. P. (1983). *Biochem. J.*, **216**, 27–36.

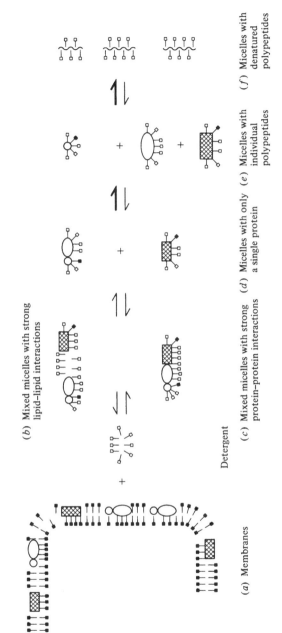

(a) Membranes

(b) Mixed micelles with strong lipid–lipid interactions

(c) Mixed micelles with strong protein–protein interactions

(d) Micelles with only a single protein

(e) Micelles with individual polypeptides

(f) Micelles with denatured polypeptides

Detergent

Fig. 4.11. The complex physical states existing on solubilization of membranes using an excess of detergent. Detergent denoted □, phospholipid ■. Reprinted from Newby, A. C., Chrambach, A. & Bailyes, E. M. (1982). In *Techniques in Lipid and Membrane Biochemistry*, pp. 1–22 (Chapter B409), *Vol. B4I1*, (edited by H. L. Kornberg & K. F. Tipton). Elsevier/North-Holland: Amsterdam.

common biochemical techniques of protein purification (salt precipitation, ion exchange chromatography, gel filtration and affinity chromatography; see Chapter 1).

Protein asymmetry in the plasma membrane

A wide variety of non-penetrating reagents, i.e. reagents which cannot cross the lipid bilayer, have been used to demonstrate protein asymmetry. They are usually radiolabelled or fluorescently tagged to allow sensitive detection and quantitation of labelling. Chemical reagents used include FMMP, which reacts with amino groups and is therefore able to label aminophospholipids such as PE (see above) as well as membrane proteins. A wide variety of other reagents capable of labelling specific groups of amino acids have been used as well as procedures such as lactoperoxidase catalysed H_2O_2-dependent iodination (using radioactive ^{125}I) of exposed tyrosine residues. Lactoperoxidase, being a large protein, cannot penetrate the phospholipid bilayer, restricting iodination to proteins at the external surface of the membrane. Proteases can also be used to attack membrane proteins from the outer or both surfaces. All these reagents require a means of detecting the protein of interest which may simply require SDS–PAGE or sometimes more specific methods such as reaction with a specific antibody. They also require the availability of sealed and broken purified membrane systems, which is simple in the case of red blood cells but more difficult and requiring subcellular fractionation techniques (see below) for other mammalian cells. Despite the difficulties, some very elegant means of showing the transmembrane nature of particular proteins in nucleated cell plasma membranes have been described and many integral membrane proteins have been categorized into the asymmetric types shown in Fig. 4.9, all molecules of the same protein having the same membrane symmetry.

One class of proteins in the plasma membrane requires special mention in terms of asymmetry. These proteins are glycoproteins containing mono- or oligo-saccharide moieties covalently linked to the polypeptide chain. The carbohydrate is present only on the extracellular face of the plasma membrane and can play an important role in the antigenic properties of the cell surface. Some of the cell surface glycoproteins have enzymic activity with the active site of the enzyme located on the outer surface of the cell. Such enzymes are called ectoenzymes and include the phosphohydrolases alkaline phosphatase and 5'-nucleotidase which are clinically useful in diagnosis (see Chapter 3). The functions of cell surface ectoenzymes are largely unknown, though involvement in nutrition, scavenging and cell–cell interactions has been suggested.

Electron microscopy and the realization of the universal bilayer structure of biological membranes

The phospholipid bilayer as the fundamental structure of the plasma membrane was proposed largely from the studies on the red blood cell described above. However, despite the Davson–Danielli model, investigation of the physical structure of the plasma membrane was hampered, since even when isolated, the thickness of the membrane (approx. 7.5–10.0 nm) was beyond the resolution of the light microscope. In the 1940s, the development of two techniques, electron microscopy and subcellular fractionation, led to a great increase in our knowledge of membranes.

The resolution of the light microscope is limited by the wavelength of light, whereas in the electron microscope the greater the speed of the electrons, the shorter their wavelength. Technical difficulties in focussing electron beams and handling biological specimens slowed the achievement of the full potential of the technique, but even with early machines, the wonderful array of cytoplasmic structures and membranous systems in mammalian cells soon became apparent (Figs 4.12 and 4.13). The double membranes surrounding the nucleus and mitochondrion were described along with the extensive intracellular reticular membrane systems comprising the endoplasmic reticulum and Golgi apparatus plus the membranes surrounding newly found organelles such as lysosomes, peroxisomes and secretory vesicles. In all cases, at high power and with appropriate electron dense stains (e.g. OsO_4), the cross-section of the membrane could be observed as an electron-dense tramline, suggesting the universality of this motif as the basis of membrane structure. Indeed, the model of membrane structure consisting of phospholipid bilayer plus protein is now accepted for membranes varying from the mitochondrial inner membrane (75 % protein by weight) to the myelin sheath (25 % protein by weight), a highly specialized form of the plasma membrane responsible for insulating myelinated nerves.

Consideration of the manipulations necessary to produce an electron micrograph by transmission electron microscopy (TEM) makes it seem remarkable that the technique has provided so much information about the membranous structures in living cells. The tissue must be fixed (often with reagents such as glutaraldehyde to cross-link proteins), stained with electron-dense heavy metals (OsO_4, $KMnO_4$, uranyl acetate), dehydrated and embedded in plastic. Then, very thin sections are cut (due to the small depth of focus of the electron beam), placed in a vacuum and bombarded by a high-energy electron beam. Photographs are then taken and developed, so that the eventual image studied by the observer, which consists of silver grains on paper, is far removed from the original

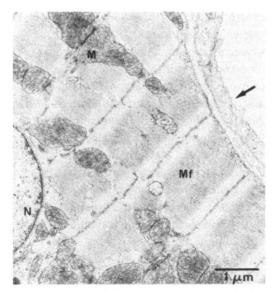

Fig. 4.12. Transmission electron micrograph of a rat heart muscle cell *in situ*. N, nucleus; M, mitochondrion; Mf, myofilament. The arrow shows an endothelial cell lining an adjacent blood capillary. Micrograph kindly given by Dr Kathryn Howell.

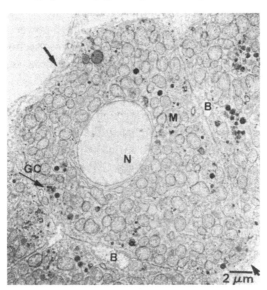

Fig. 4.13. Transmission electron micrograph of a rat liver hepatocyte *in situ*. N, nucleus; M, mitochondrion; GC, Golgi complex; B, bile canaliculus. Note the prominence of intracellular membranes. In this cell type, the plasma membrane accounts for 2% of total cell membrane. The bold arrows point to adjacent blood sinusoid lining cells. Micrograph kindly given by Dr Kathryn Howell.

living cell. Nonetheless, many observations suggest the essential correctness of the images provided by TEM. First, it is possible to relate the overall dimensions and major features of a living cell observed under the light microcsope to the images from TEM. Secondly, it is possible to show the same localization of specific macromolecules in the cell using light microscopy or TEM. For example, the stained products of some enzyme reactions can be observed in the light microscope in fixed sections and compared with the data from TEM when one uses an electron-dense stain. Many phosphatases which hydrolyse specific organic phosphates to release inorganic phosphate have been localized in this way by adding soluble lead nitrate to form insoluble lead phosphate precipitates at the site of the enzyme activity. More generally, antibodies to specific macromolecules, labelled with fluorescent probes for light microscopy or electron-dense probes for TEM can be used to show localization to specific parts of the cell (Fig. 4.14). Thirdly, it is possible to isolate specific parts of the cell by subcellular fractionation techniques (see below), showing that they retain the TEM structure and specific enzymes predicted from TEM histochemical studies. Finally, it has been possible to correlate TEM information about the structure of organelles such as the plasma membrane and ribosomes with information provided by other physical techniques such as X-ray diffraction, nuclear magnetic resonance and polarizing light microscopy.

Since the introduction of TEM, a number of other electron micro-

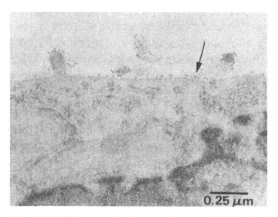

Fig. 4.14. The use of electron-dense gold-labelled antibodies to localize cell surface components. Anti-(plasma membrane) antibodies were bound to cultured rat liver cells at 0 °C, excess antibody washed off and 5 nm colloidal gold-protein A complex added. This bound specifically to the antibodies and the cells were then fixed and processed for TEM. Micrograph kindly given by Dr Kathryn Howell.

(a)

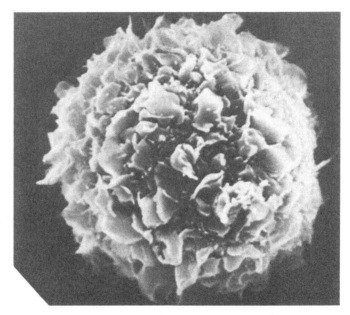

(b)

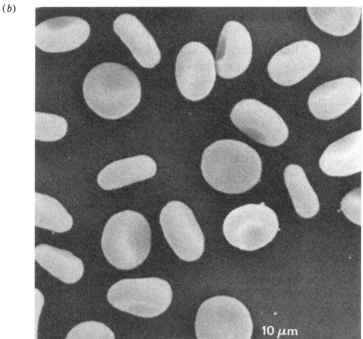

10 μm

Fig. 4.15. Scanning electron micrograph of the cell surface. Different cell types have very distinctive cell surface projections when observed by SEM. (*a*) Cell isolated from rat liver, diameter approximately 15 μm. Micrograph kindly given by Prof. Lawrence Herman. (*b*) Mammalian erythrocytes showing smooth surface and biconcave shape. Micrograph kindly given by Dr Janet Stein.

scopic techniques have been introduced which have lead to a greater knowledge of plasma membrane structure. Scanning electron microscopy (SEM) produces an image of the cell surface showing structures such as microvill and other surface projections, thus allowing visualization of the variety of surface morphology on different cell types (Fig. 4.15). Freeze–fracture techniques have helped to show the organization of proteins within the lipid bilayer. In these methods, specimens are frozen (−196 °C) and fractured under vacuum. A small amount of ice is allowed to sublime from the fractured surface and the etched surface is then shadowed with platinum and carbon, the organic material digested away and the platinum replica examined in the TEM. The important point is that, whereas fractures may occur along the outside surfaces of the membrane, they are favoured between the two halves of the bilayer. This was shown in experiments using labelled model bilayers (Fig. 4.16). 'Bumps' in the images seen in TEM of freeze–fractured plasma membranes are thought to be integral membrane proteins since artificial phospholipid bilayers show no 'bumps' unless proteins are incorporated. The technique of freeze–fracture has provided further evidence that the basic structure of the plasma membrane is a phospholipid bilayer.

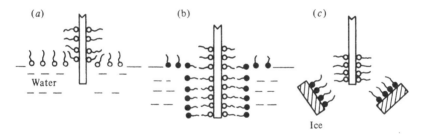

Fig. 4.16. The use of radioactively labelled lipid bilayers to show that freeze–fracture splits a membrane through the centre of the bilayer. An unlabelled lipid monolayer is first formed on a glass cover slip (*a*) and the coverslip subsequently dipped into a radioactively labelled lipid monolayer to form a bilayer (*b*). The bilayer is split by freeze–fracture, radioactivity remaining with the ice (*c*).

Subcellular fractionation and the preparative isolation of membranes

The analysis of the composition and function of membranes from nucleated mammalian cells requires the ability to isolate pure fractions of these membranes in reasonable quantities. The major techniques for doing this rely on a variety of methods of disrupting tissues and cells followed by the use of ultracentrifugation methods. Usually, cells and tissues are disrupted by suspending in isotonic medium (0.25 M sucrose buffered to pH 7.4) and homogenizing using glass/glass or teflon/glass homogenizers (Fig. 4.17), although more esoteric methods including gas cavitation, tissue presses and sonication are sometimes used. In general, the aim of the experimenter is to choose an homogenization method that avoids damaging the membrane structure of the organelles of interest.

Ultracentrifuges can be used to fractionate homogenates in essentially two ways. In the first, known as *differential centrifugation*, the homogenate is rotated at different speeds to provide pellets that are found to contain different subcellular elements. Thus, if a liver homogenate is centrifuged at 1000 g for 10 min, the pellet is rich in nuclei; when the 1000 g supernatant is spun at 10 000 g for 10 min, the pellet is rich in mitochondria; when the 10 000 g supernatant is spun at 100 000 g for 1 h, the pellet is rich in microsomes (the name given to the membrane vesicles formed as an artefact by disrupting the endoplasmic reticulum). The theoretical reason for the separation achieved is that the rate at which a component sediments is dependent on its size and shape and can be expressed as its sedimentation coefficient (defined as the velocity with which a particle moves per unit acceleration); thus, large nuclei are easily separated from small microsomes. However, many organelles are polydisperse after homogenization: e.g. plasma membrane may be dis-

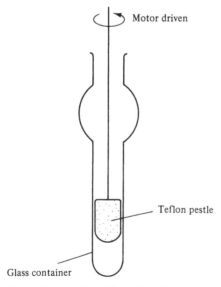

Fig. 4.17. A teflon/glass tissue homogenizer.

rupted into fragments of different size and shape and in the scheme described above would be found in the 1000 g 'nuclear' pellet, 10 000 g 'mitochondrial' pellet and 100 000 g 'microsomal' pellet fairly evenly divided between them. In addition, particles starting near the bottom of the homogenate will enter the pellet before particles starting near the top, even when the latter have higher sedimentation coefficients.

To overcome the difficulties with differential centrifugation, *density gradient centrifugation* was introduced. A preformed density gradient is prepared, for example by careful filling of a centrifuge tube with a solution of decreasing sucrose concentration, and the homogenate or a fraction from differential centrifugation loaded on to the top and centrifuged at high speed (usually >10 000 g) in a swing-out rotor. Density gradients may be used to achieve 'rate separations', where centrifugation is carried out for relatively short periods (e.g. 1–2 h) and separation is dependent largely on the size and shape of particles as in differential centrifugation. Alternatively, centrifugation continues until equilibrium conditions are achieved (often 18–24 h) and cellular components are separated on the basis of their buoyant density (dependent on organelle contents and membrane composition) rather than their size. Such separations are termed 'equilibrium or isopycnic separations'. In both applications of density gradient centrifugation, the sucrose gradient prevents convective mixing of the subcellular components. It is not itself sufficiently dense for the gradient to alter during centrifugation (cf. CsCl gradients, Chapter 7). Many variations of density gradient centrifugation exist using different density gradient media (often of higher molecular weight than sucrose to avoid the potentially damaging osmotic effects of high sucrose concentration) and a variety of differently designed centrifuge rotors.

Clearly, a problem in subcellular fractionation is the identification of fractions of interest and the definition of their purity. Although electron microscopy plays an important role in this, it is slow and too cumbersome for the routine analysis of many fractions. De Duve, a Belgian scientist solved the problem by using biochemical markers, particularly marker enzymes to identify fractions. He made two assumptions:

1. The postulate of biochemical homogeneity, stating that members of a given particle population have the same biochemical composition, larger particles simply having more of everything than smaller ones.
2. The postulate of single location, that each enzyme (or marker) is restricted to a single intracellular site.

Thus, enzymes such as cytochrome oxidase can be used to define mitochondria, 5'-nucleotidase the plasma membrane, acid phosphatase the lysosome, glucose-6-phosphatase the liver microsomes, and markers such as DNA the nucleus. Moreover, since enzymes are all proteins, a purer fraction will have a higher specific activity (activity per mg protein) of its marker enzyme than the homogenate or other fractions. It is therefore simple to follow the purification of a subcellular fraction and, by measuring the amount of marker enzyme and keeping a balance sheet, determine its yield as well as purification.

The power of density gradient centrifugation to prepare membrane fractions is demonstrated by the ability to isolate membrane systems such as the Golgi complex (Fig. 4.18), to separate inner and outer mitochondrial membranes and to separate domains of the plasma membrane (see below).

It is difficult to overestimate the impact of subcellular fractionation techniques to modern biochemistry, especially since, not long ago, many biological scientists would have echoed the views of an eminent Cambridge histologist who often claimed that if a biochemist wanted to know what was going on in a specific room in a particular building, he would first smash down the building then measure the juices coming out of the drains. The advent of the techniques of subcellular fractionation at least

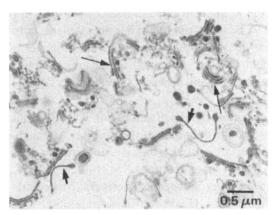

Fig. 4.18. Transmission electron micrograph of a Golgi fraction isolated from rat liver. The fraction contains reasonably intact Golgi stacks (long arrows), some single cisternae and numerous membrane vesicles. The fraction is not absolutely pure and many of the membrane vesicles are not of Golgi origin but come from other smooth membrane systems in the cell, e.g. plasma membrane and smooth endoplasmic reticulum. Micrograph kindly given by Dr Kathryn Howell.

allowed the possibility of cataloguing and analysing the bricks and mortar of the cell.

The fluid mosaic model of the plasma membrane

By the late 1960s, sufficient information was available on a large number of cell membranes to require a considerable revision of the models of membrane structure. As described earlier (Fig. 4.9), it was apparent that many membrane proteins were integral proteins with hydrophobic regions interacting with the interior of the phospholipid bilayer. From thermodynamic considerations of such interactions, the fluid mosaic model of membrane structure (Fig. 4.19) was proposed by Singer and Nicholson and gave a considerable fillip to our understanding of membrane dynamics. Two proposals were made. The first of these was that the lipid and protein molecules were arranged in a tightly packed, water-excluding mosaic. Thus, the apolar regions of the proteins and lipids interacted in the hydrophobic membrane core wth hydrophilic regions of each amphipathic molecule solvated in the water outside. Secondly, it was proposed that individual membrane components are free to move in the plane of the membrane.

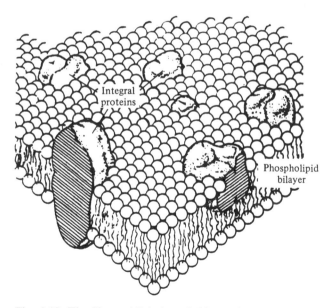

Fig. 4.19. The Singer–Nicholson fluid mosaic model of membrane structure. The model shows globular integral proteins (with stippled surfaces) randomly distributed in the plane of the phospholipid bilayer.

The overall structure of the fluid mosaic model was supported by a number of experimental approaches, not least the classical data suggesting a phospholipid bilayer and evidence for the existence of integral and peripheral membrane proteins. In addition, freeze–fracture studies had suggested that integral proteins were not packed closely together in the membrane and this agreed with considerations of the amount of lipid and protein in a membrane. While these components may be equal by weight, the molecular weight of a phospholipid (a few hundred) is very much less than that of a membrane protein (e.g. 50 000), allowing tens of lipid molecules per protein molecule, and leading to the idea of biological membranes being two dimensional 'seas' of lipid in which the integral membrane proteins float.

Lateral diffusion of phospholipids

The study of phospholipid diffusion commenced with artificial phospholipid systems, particularly spherical vesicles called liposomes. These are made simply by suspending phospholipids in aqueous solutions where they spontaneously form thermodynamically stable bilayers. Lipid molecules, chemically constructed such that the polar head group carries a 'spin-label' (e.g. a nitroxyl group), can be inserted into such a bilayer or a natural membrane and the 'spin-label' used as a reporter of the lipid molecule's movement. The 'spin-label' contains an unpaired electron whose spin creates a paramagnetic signal detectable by electron spin-resonance spectroscopy (Esr). It was found that, in both model bilayer systems and biological membranes, lipid molecules diffuse laterally with a rapid diffusion coefficient of $10^{-8} \, cm^2 \, s^{-1}$. Thus, it would take only a few seconds for a lipid molecule to travel the length of the cell surface of a mammalian cell. In contrast, Esr showed that phospholipids have a very slow rate of 'flip flop' between one half of the bilayer and the other with a half-life of more than 6 h.

Lateral diffusion of proteins

Even more dramatic data, produced in the early 1970s, showed that integral membrane proteins also diffuse laterally. A classic experiment performed by Frye and Edidin investigated the redistribution of cell surface antigens after fusion of cultured mouse and human cells (Fig. 4.20). Each cell surface could be labelled using specific antibodies tagged with a particular fluorescent label, fluorescein (green)-tagged antibodies to mouse cells and rhodamine (red)-tagged antibodies to human cells. The cells were fused with inactivated Sendai virus to form heterokaryons (mouse–human cell hybrids) which initially had large surface areas

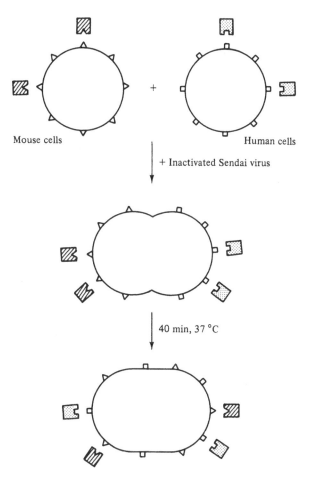

Mouse cells   Human cells

+ Inactivated Sendai virus

40 min, 37 °C

Fig. 4.20. The redistribution of surface antigens in fused cells detected with fluorescent-labelled antibodies. The stippled and hatched symbols represent antibodies labelled with distinguishable fluorescent markers.

capable of being labelled by the red or green antibodies, but after 40 min at 37 °C showed complete intermixing of the antigens labelled by the two antibodies. Intermixing of the cell surface antigens was not affected by inhibitors of ATP synthesis or protein synthesis and was consistent with the lateral diffusion of integral membrane proteins in the phospholipid bilayer.

Other experiments also suggested lateral diffusion of integral membrane proteins. In particular, antibodies added to lymphocytes were shown to cause the redistribution of surface antigens into patches and eventually caps at one end of the cell, causing eventual endocytosis and uptake into the cell. In these experiments, the antibodies were fluorescently tagged for visualization in the light microscope or ferritin-tagged for visualization in the electron microscope. Since polyclonal antibodies consist of a mixture of divalent (and/or multivalent) antibodies directed at different antigenic sites on the same protein, it is possible for them to form a lattice of antigenic proteins in the membrane by cross-linking. The resultant patch, a raft of proteins in the phospholipid bilayer, can grow and be swept along to form a cap at one end of the cell. Univalent Fab fragments are ineffective in patching.

Measurement of the movement of labelled antibodies on the cell surface provided values of the diffusion coefficients of integral membrane proteins which were supported by experiments using properties of integral proteins themselves. For example, membranes containing rhodopsin molecules (the visual pigment which contains the chromophore 11-*cis*-retinal covalently attached to the protein, and con-stitutes 50 % by weight of the retinal disc membrane) could be photobleached by laser light and the rate of diffusion of unbleached rhodopsin molecules back into the bleached area measured. In general, diffusion coefficients for integral membrane proteins are less than those measured for phospholipids, being in the range 10^{-9}–10^{-19} cm^2 s^{-1}.

Membrane domains and the interaction of membrane components with cytoskeletal proteins

Light microscopic examination of mammalian tissues shows that many cells are polar with specific cell surface modifications at one end of the cell. This is particularly clear in cases such as columnar epithelial cells (enterocytes) lining the gut, but is also true for cells such as parenchymal liver cells (hepatocytes), which separate the body fluids blood and bile (Fig. 4.21). These cells have clearly defined plasma membrane domains at different parts of the cell surface, which differ in their complement of membrane enzymes, receptors and overall protein/phospholipid com-position. Just as it is possible to separate subcellular organelles using the techniques of subcellular fractionation, it is also possible to separate and isolate these membrane domains since they differ in density as a result of their different composition. In the case of hepatocytes, one can isolate the blood sinusoidal domain (characterized by the presence of the enzyme adenylate cyclase and many hormone receptors) from the con-tinguous domain (characterized by specific structures involved in cell–cell contact: tight junctions, gap junctions and desmosomes), and from the bile canalicular domain (characterized by the highest concentrations of

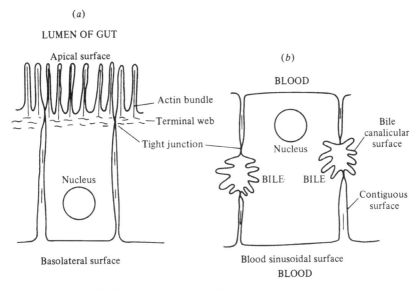

Fig. 4.21. Schematic views of the cell surface domains of enterocytes (*a*) and hepatocytes (*b*). The apical surface of an enterocyte has extensive microvilli, each of which contains an actin filament bundle. These microvilli greatly increase the cell surface available for the absorption of nutrients from the gut.

certain cellular phosphatases such as 5′-nucleotidase and alkaline phosphatase).

The existence of membrane domains appears to contradict the simplistic view of the fluid membrane, but may be explained as a result of diversification of the basic fluid mosaic structure. Thus, the mobility of individual membrane proteins may be affected by difference in fluidity between the two halves of the bilayer, protein–protein or protein–lipid interactions, sequestration or exclusion from particular lipid domains, restraint by peripheral membrane proteins or restraint by membrane-associated cytoskeletal elements at the cytoplasmic surface. There is increasing evidence that the domain structure of the surface of nucleated cells may be the result of interactions between membrane proteins and the cytoskeleton, but the best existing description of such interactions is for the red blood cell (Fig. 4.22), in which there is no evidence for other than a uniform composition of the plasma membrane over the entire cell surface.

The mammalian red blood cell has a simple cytoskeleton based on the proteins actin and spectrin, and has no microtubules or intermediate filaments. Most of the actin is present in the filamentous F-form and

(a)

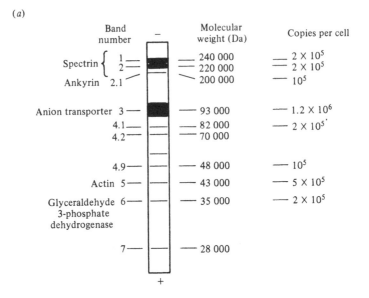

Band number	Molecular weight (Da)	Copies per cell

Spectrin { 1 — 240 000 — 2×10^5
2 — 220 000 — 2×10^5
Ankyrin 2.1 — 200 000 — 10^5

Anion transporter 3 — 93 000 — 1.2×10^6
4.1 — 82 000 — $2 \times 10^{5'}$
4.2 — 70 000

4.9 — 48 000 — 10^5
Actin 5 — 43 000 — 5×10^5
Glyceraldehyde 6 — 35 000 — 2×10^5
3-phosphate
dehydrogenase

7 — 28 000

(b)

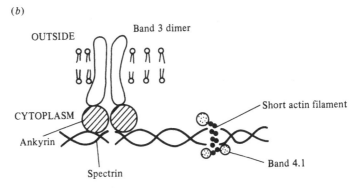

Fig. 4.22. Proteins of the red blood cell membrane. (*a*) Diagram of SDS–PAGE separation of the major proteins of the human red blood cell membrane. (*b*) The interaction of Band 3 protein with the red blood cell cytoskeleton.

oligomers of 10–17 molecules are cross-linked by spectrin. Spectrin composes 60–70 % by weight of the cytoskeleton and consists of two similar but not identical subunits (Mr 240 000 and 220 000, respectively), with structural homologies to the muscle protein myosin, organized as filamentous heterodimers. Immunoelectron microscopic localization of spectrin in the red blood cell has shown it to be concentrated on the cytoplasmic side of the plasma membrane. A further protein, ankyrin, is involved in this cytoskeletal framework and links spectrin heterodimers to the transmembrane, anionic transport protein, Band 3. These interactions have been shown using chemical cross-linking reagents. While Band 3 is shown thus to anchor the plasma membrane to the cytoskeleton, only about 10–20 % of all Band 3 molecules are involved in this process. These Band 3 molecules are linked to a peripheral protein, Band 4.1, which may play a role in the plasma membrane–cytoskeleton interactions. Non-erythroid ankyrin and spectrin with homologies to the erythrocyte forms have been found in nucleated cells, suggesting a general function for these molecules in plasma membrane–cytoskeleton linkage, though this has not been established.

Nucleated mammalian cells have several cell surface microdomains involved in cell–cell interactions (Fig. 4.23). These microdomains are named *tight junctions*, where the plasma membrane of two cells are tightly apposed, *gap junctions* (nexuses), which contain arrays of specific gap junction transmembrane proteins, *spot desmosomes* (macula adherens) and *belt desmosomes* (zonula adherens and the closely related fascia adherens). Tight junctions are particularly important in epithelial cells, e.g. playing an important role in intestinal epithelium separating the contents of the gut lumen from blood and lymph and also keeping the membrane proteins of the apical surface separate from those of the basolateral surface (Fig. 4.21). Gap junctions allow ion and small molecule movement between a cell and its neighbours, whereas spot desmosomes act as spot welds to hold adjacent cells together. There is good microscopic evidence suggesting a close apposition of intermediate filaments (keratin) with spot desmosomes and actin filaments with belt desmosomes. Despite considerable efforts, the biochemical mechanisms of interaction between cytoskeletal elements and integral plasma membrane proteins remain obscure.

Membrane synthesis and turnover

The plasma membrane is not a static structure and its macromolecular components are covalently replaced as a result of synthesis and degradation. Within the eukaryotic cell, membrane synthesis occurs in the endoplasmic reticulum. This contains the enzymes responsible for phospholipid and cholesterol synthesis, although, as discussed below, most human cells obtain cholesterol from low-density lipoprotein in the bloodstream. Phospholipid synthesis (Fig. 4.24) results in the incorporation of new phospholipids into the cytoplasmic half of the phospholipid bilayer. These must then be redistributed across the bilayer. While such 'flip flop' occurs only very slowly in the plasma membrane, it appears to

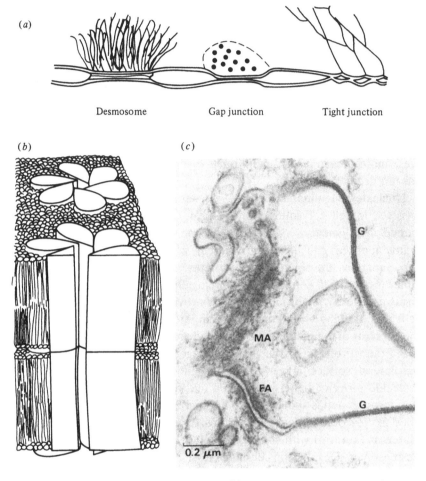

Desmosome   Gap junction   Tight junction

Fig. 4.23. Cell–cell interactions. (*a*) Diagrammatic representation of interactions between cell surfaces. (*b*) Model of gap junction structure showing hydrophilic transmembrane channels composed of six polypeptide subunits in register forming intercellular pores. (*c*) Transmission electron micrograph of cardiac muscle preparation showing macula adherens (MA), fascia adherens (FA) and gap junction (G). Gap junction model and micrograph kindly given by Dr Camilo Colaco.

be catalysed in the endoplasmic reticulum, though the mechanism is not fully understood. Endoplasmic reticulum phospholipids may move with newly synthesized proteins to other organelles. However, their movement may also be facilitated by phospholipid exchange proteins, small (Mr 14 000), water-soluble, basic proteins found in the cytosol.

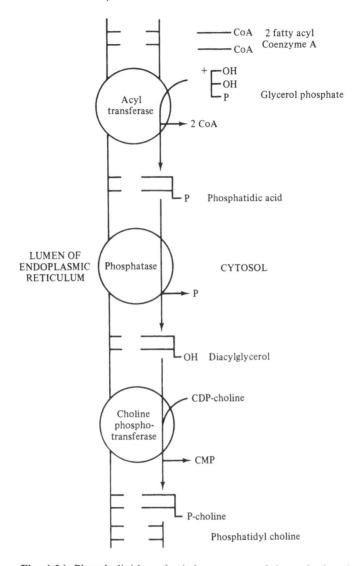

Fig. 4.24. Phospholipid synthesis by enzymes of the endoplasmic reticulum. The synthesis of phosphatidyl choline (PC) is shown. Note the function of cytidine diphosphate (CDP) in attaching polar head groups to form newly synthesized phospholipid. Phosphate groups are indicated by the symbol P.

The endoplasmic reticulum is also responsible for the synthesis of plasma membrane proteins. It is now known that integral membrane proteins are made by a modification of the method used to synthesize secreted proteins (Chapter 6) which has been explained by the signal hypothesis (Fig. 4.25). Messenger RNA (mRNA) coding for a membrane protein passes from the nucleus to the cytoplasm where ribosomes bind and commence translation. This occurs in the same way as for all other proteins, the mRNA being read in the 5′ to 3′ direction and the new polypeptide chain being synthesized N-terminal to C-terminal. If the polypeptide being made is a cytosolic protein or peripheral membrane protein, synthesis will proceed to completion. However, if the polypeptide chain is an integral membrane protein or secreted protein (see also Chapter 6), the first N-terminal section of sequence (approx. 15–30 amino acids, predominantly hydrophobic) forms a signal sequence which attaches the mRNA–ribosome–polypeptide complex to the endoplasmic reticulum (hence the characteristic electron microscopic appearance of rough endoplasmic reticulum, rich in bound ribosomes). After binding to the endoplasmic reticulum, protein synthesis will proceed with cotranslational insertion of the protein through the endoplasmic reticulum membrane. A signal peptidase in the lumen of the endoplasmic reticulum cleaves the signal peptide and the fully synthesized polypeptide is either completely transferred into the lumen in the case of a secreted protein or remains as a transmembrane protein. The latter requires the existence in the polypeptide chain of a stop–transfer signal (analogous to the signal peptide being a start–transfer signal), and examination of the amino acid sequence of a number of purified membrane proteins suggests that the stop signal includes charged

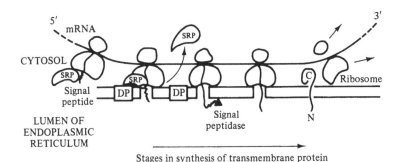

Stages in synthesis of transmembrane protein

Fig. 4.25. The signal hypothesis for synthesis of a transmembrane protein. Synthesis of the protein commences in the cytosol but is arrested by the signal recognition particle (SRP). This binds to a docking protein (DP) in the endoplasmic reticulum membrane and is released once the signal peptide inserts in the membrane allowing synthesis to resume. An integral protein traversing the membrane once is shown. Further description of the signal hypothesis is given in Chapter 6.

amino acids that would not be thermodynamically favoured for crossing the phospholipid bilayer. The mechanism described for cotranslational insertion of a new polypeptide chain into the endoplasmic reticulum membrane requires modification for integral membrane proteins crossing the bilayer more than once. Evidence for the necessary internal start–transfer sequences has been obtained, and indeed such sequences are also essential for those mitochondrial membrane proteins synthesized on cytoplasmic ribosomes and released into the cytosol before incorporation into the mitochondrial membrane. The cotranslational insertion mechanism also requires modification for integral membrane proteins that have their amino-terminus on the cytoplasmic side of the membrane. In this case, it has been suggested that the initially synthesized N-terminal portion of the polypeptide chain forms a hairpin loop in the endoplasmic reticulum membrane, allowing the remainder of the polypeptide and C-terminus to pass across the membrane into the lumen.

Much of the evidence for cotranslational insertion of integral membrane proteins into the endoplasmic reticulum came initially from studies on the synthesis of viral coat proteins. After infection of a mammalian cell, coated viruses such as vesicular stomatis virus and semliki forest virus expropriate the cell's protein synthesis machinery so that the cell produces large amounts of the proteins which together with phospholipid form the external coat of newly synthesized virus. When mRNA coding for viral coat proteins is added to a cell-free translation system containing all the cytoplasmic factors for protein synthesis, the coat protein produced can be inserted cotranslationally into the membrane of added smooth microsomes, but cannot be inserted if the microsomes are added after synthesis is complete.

Investigation of the passage of newly synthesized viral coat proteins from the endoplasmic reticulum via the Golgi apparatus to the plasma membrane, from where new virus buds off, has demonstrated the route taken by plasma membrane proteins and also helped to elucidate the biochemical mechanisms of glycosylation to which these proteins are subject. Many plasma membrane proteins are glycosylated, the sugar residues being on the outside of the cell. Carbohydrate side-chains may be attached by an oxygen ester linkage to hydroxyl containing amino acids such as serine and threonine (O-linked sugars) or through a nitrogen ester to an asparagine residue (N-linked sugars). N-linked sugars are added by a complex mechanism involving core glycosylation of the newly synthesized membrane protein in the endoplasmic reticulum with subsequent trimming and elaboration in both the endoplasmic reticulum and Golgi apparatus. Whereas the carbohydrate chains are built up (and O-linkage provided) using nucleotide-sugars, specifically UDP-sugars as substrates, attachment of oligosaccharide by N-linkage requires the prior formation of a complex oligosaccharide linked to the lipid dolichol. The lipid linkage appears to ensure that the sugars are well placed on the face of the membrane to glycosylate appropriate membrane proteins. Since the steps involved in glycosylating plasma membrane proteins are sequential and occur as the protein passes through the endoplasmic reticulum and Golgi apparatus, it is possible to identify the stage in the transfer pathway reached, by the glycosylation state of the protein. Particularly elegant experiments have been devised mixing Golgi fractions from one cell mutant lacking a terminal glycosylating enzyme with Golgi fractions from a cell containing the enzyme. Viral coat protein produced in the mutant cell can complete its glycosylation only as a result of transfer in small vesicles to the Golgi apparatus containing the enzyme. This type of experiment is being exploited to establish the mechanisms of movement of newly synthesized plasma membrane proteins from the endoplasmic reticulum, through the Golgi apparatus to the plasma membrane.

One of the major remaining problems in understanding the synthesis of plasma membrane proteins is the mechanism of sorting. It is not clear why some newly synthesized membrane proteins remain in the endoplasmic reticulum, others travel as far as the Golgi apparatus and yet others move to the plasma membrane. Coated viruses may again help in terms of resolving the way in which plasma membrane proteins are separated into cell surface domains, since it is now possible to study the intracellular synthesis of the coat proteins of different viruses which are inserted at different ends of the cell.

The mechanism of sorting is also a difficulty when a plasma membrane protein is to be degraded. Most studies have indicated that plasma membrane proteins are relatively long lived, with half-lives of 2–5 days, compared to cytoplasmic proteins which, in exceptional cases, can have a half-life as short as 10 min. The major mechanism of degradation is thought to be internalization and degradation in lysosomes, although shedding of some membrane components by cells has also been reported. It is not known if any specific membrane damage event can trigger the degradation of a plasma membrane protein, though some membrane proteins are degraded at a greater rate when internalization is triggered.

Endocytosis and the recycling of membrane components

The dynamic nature of the cell surface plasma membrane is now known to involve not simply two dimensional fluidity nor just synthesis and degradation of macromolecules, but also the constant internalization and recycling of its components. The process of plasma membrane internalization is known as endocytosis and is the only physiological route by which macromolecules and small molecules which are neither lipid-soluble nor have transport mechanisms, can gain access to cells. It involves invagination of the plasma membrane to form an endocytic vesicle containing the extracellular material. Endocytosis may be conveniently subdivided into phagocytosis (eating), where large particles visible by light microscopy are taken up as a result of close apposition of the plasma membrane to the particle, and pinocytosis (drinking), which describes the vesicular uptake of everything from small particles to soluble macromolecules and low molecular weight solutes. Whereas phagocytosis is particularly pronounced in specialized cells such as macrophages and polymorphonuclear leucocytes responsible for clearing tissues of invading bacteria and other debris, pinocytosis is a feature of all mammalian cells. Endocytosis can be further subdivided into fluid phase endocytosis, where uptake is proportional to the concentration of solute in the extracellular fluid, and receptor-mediated endocytosis (or adsorptive endocytosis), where uptake in addition depends on the number, affinity and function of cell surface binding sites. Receptor-mediated endocytosis is therefore selective and can concentrate large amounts of a specific solute without ingesting a correspondingly large volume of solution.

Although endocytosis performs important nutritive functions for the cell, the process identifies a problem for cell biologists interested in the plasma membrane. For example, experiments using quantitative electron microscopy have shown that, as a result of pinocytic activity, cultured fibroblasts and macrophages internalize 50–200 % of their surface area per hour (i.e. much faster than the rate of turnover – synthesis and degradation – of individual proteins and lipids) without alteration in cell volume or surface area. The problem of plasma membrane internalization with maintenance of cell surface area is not confined to phagocytosis and pinocytosis, since cells involved in secretion (Chapter 6) must internalize secretory vesicle membrane after fusion of secretory vesicles with the plasma membrane if the cell is not to expand.

During the 1970s, a number of experimental approaches showed that it was possible for membrane components to be internalized and recycled back to the cell surface. One set of experiments from de Duve's institute showed this with particular elegance by allowing the uptake of labelled anti-IgG antibodies into the lysosomes of cultured cells and then measuring the reappearance of these antibodies at the cell surface coupled to recycling anti-plasma membrane antibodies (Fig. 4.26). The experiments also provided evidence that endocytosed plasma membrane components

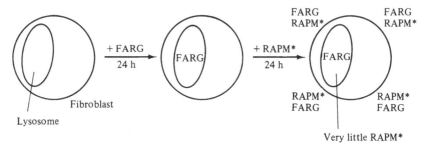

Fig. 4.26. An experiment using labelled antibodies to demonstrate
the recycling of plasma membrane components in rat embryo
fibroblasts. The cells were incubated with fluorescent-labelled goat
anti-rabbit immunoglobulin (FARG) which was endocytosed and
loaded the lysosomes. Subsequent incubation with radioactively
labelled rabbit anti-(fibroblast plasma membrane) antibodies
(RAPM*) labelled the cell surface. After incubation for 24 h, cell
surface RAPM* was found to be associated with FARG as a result
of membrane cycling. When the experiment was carried out with
radioactively labelled control rabbit immunoglobulin instead of
RAPM*, these antibodies simply loaded the lysosomes along with
FARG.

recycle via the lysosome, though this is not thought to be a general route
and there is evidence that, in different systems, recycling can occur via the
lysosome, lysosome and Golgi apparatus or directly from an organelle
termed the endosome that acts as the initial repository of internalized
plasma membrane components.

It is now known that the site of receptor-mediated endocytosis is at
characteristic indentations of the plasma membrane known as coated
pits. In the electron microscope, these have a fuzzy coating on the
cytoplasmic side and it is thought that invagination of the membrane
occurs at these coated pits to form coated vesicles. Such vesicles have
been isolated from many tissues and the coat analysed and found to
consist mainly of a protein of Mr 180 000 called clathrin and shown to be
capable of forming regular closed-basket structures on the cytoplasmic
membrane surface of a coated vesicle. Other minor proteins are also
present in the protein coat and may be responsible for targeting coated
vesicles to particular sites in the cell or even triggering the coating/
uncoating process under physiological conditions. It has been suggested
that coated pits may be the sites of all pinocytic events at the cell surface,
including fluid phase endocytosis, though this remains an area of conten-
tion. In addition, it has been suggested that coated pits act as molecular
filters at the cell surface, somehow allowing the internalization of some
plasma membrane molecules but not others.

For some time it was thought that internalized coated vesicles lost their coats and fused with the lysosome, but careful electron microscopic examination of cells after internalizing electron-dense labelled ligands has shown that they usually fuse with the endosome, an extensive intracellular membranous structure (Fig. 4.27). This appears to consist of reticular and vesicular elements which may have structural and functional differences at different sites in the cell. Thus, some workers describe endosomal elements close to the cell surface as differing from deep-lying elements. The function of the endosome system appears to be to separate surface-bound ligands from their receptors, allowing the latter to recycle directly to the cell surface. Ligand–receptor segregation is achieved by the presence of an acid pH within the endosome, generated by a proton pumping ATPase in the endosomal membrane. The endosomal acidity has been measured by investigating the fluorescence of fluorescein-labelled ligands taken up into cultured cells. The fluorescence intensity of this label is dependent on pH and has shown that the endosomal pH is 5.0 compared to pH 4.6 for lysosomes. Endosomal pH has also been inferred from electron microscopic observations of coated virus infection of

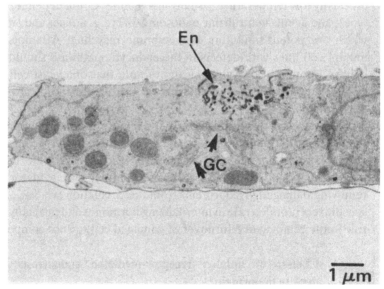

Fig. 4.27. Localization of the endocytic compartment in a cultured rat liver cell. Hepatoma cells were allowed to endocytose horse radish peroxidase for 15 min at 20 °C. They were then fixed and a histochemical staining procedure carried out to locate the peroxidase. The dense reaction product is discretely localized within the endosomal compartment (En) in a region distinct from the Golgi complex (GC). Micrograph kindly given by Dr Kathryn Howell.

cultured cells. Coated viruses such as semliki forest virus have an RNA nucleocapsid surrounded by a phospholipid bilayer membrane containing integral membrane proteins and can fuse with the plasma membrane of a target cell if the extracellular pH is below 6. Normally, the virus binds to the cell surface, is endocytosed into the endosomal compartment, where, because of the low pH, membrane fusion occurs and infectious RNA is released into the cytoplasm. Agents such as amines and ammonia that neutralize endosomal acidity prevent viral infection of the target cell.

There is still much to learn about the endosome. It has been difficult to isolate by subcellular fractionation because of the lack of suitable markers, though some partially successful attempts have been made on the basis that it is an acidic compartment (containing a membrane-bound proton pumping ATPase) into which ligands are rapidly endocytosed and which contains no lysosomal marker enzymes (e.g. acid hydrolases). The membrane composition is thought to have similarities with the plasma membrane, though the extent to which proteins are segregated between the two is uncertain. Very little is known about the mechanisms of sorting of membrane components into or out of the endosome. Conditions inhibiting endocytosis should prove useful in investigating mechanisms, especially low temperature (usually less than 15 °C, but dependent on cell type), and agents neutralizing endosomal pH (e.g. amines and ammonia) which are potent inhibitors of membrane recycling. Attempts to find mutant cell lines with defects in the endocytic pathways should also be helpful, though it is wryly amusing to note that one yeast cell mutant isolated apparently has no clathrin yet is still capable of growth and division.

It is now clear that endocytosis plays a more specific role in nutrition than previously thought, e.g. in the uptake of cholesterol (via the low-density lipoprotein receptor, see below) and iron (via the uptake of transferrin bound to its cell surface receptor), also a major role in removing damaged glycoproteins from the circulation (via various sugar specific receptors), a role in internalizing hormones and probably a major role in the removal and turnover of damaged cell surface components.

Cholesterol uptake: receptor-mediated endocytosis of low-density lipoprotein

Cholesterol is an essential component of the plasma membrane required to maintain mechanical stability. Mutant animal cell lines incapable of synthesizing cholesterol rapidly lyse unless cholesterol is added to the culture medium so that it can be incorporated into the plasma membrane. Although most mammalian cells are able to synthesize cholesterol, they preferentially obtain their cholesterol from plasma lipoproteins secreted by the liver and intestine. The great interest in cholesterol metabolism has been generated by the observation that elevated plasma cholesterol is associated with the development of atherosclerosis and coronary heart disease, the major single cause of death in the Western world (approximately a quarter of a million deaths per year in Britain alone). Careful studies by Brown and Goldstein in Dallas, Texas, have shown how lipoprotein endocytosis provides cholesterol to peripheral cells and how abnormalities in this process lead to atheroma and heart disease. At the same time, their studies have revealed patients with genetic abnormalities of lipoprotein endocytosis that have greatly increased our knowledge of the biochemistry of endocytic processes.

Man's requirement for cholesterol is about 0.35 g per day. The average diet contributes about 0.5 g per day and endogenous synthesis by the liver about 1 g, the excess being excreted as bile acids (the major cholesterol metabolite in liver) and cholesterol via the bile into faeces. Abnormalities in the excretion system can lead to cholesterol gallstones (approximately 75 % of all gallstones, and it is estimated 20 million Americans have gallstones!) when either decreased synthesis of bile salts or possibly increased secretion of cholesterol into bile occurs. The cholesterol taken up into the enterocytes and esterified, or synthesized by hepatocytes is packaged together with triacyl glycerol to make respectively chylomicrons or very low-density lipoprotein particles (VLDL) which are secreted into the bloodstream. Triacyl glycerol in these particles is hydrolysed by the enzyme lipoprotein lipase in adipose tissue causing the release and uptake into fat cells of fatty acids for re-esterification and storage. Chylomicron remnants are removed by the liver and the cholesterol ester can be hydrolysed and free cholesterol incorporated into VLDL. In the adipose tissue blood vessels, the hydrolysis of triacyl glycerol in chylomicrons and VLDL is associated with alterations in the apoproteins and, in the case of VLDL, these alterations and the action of the enzyme lecithin cholesterol acetyl transferase (LCAT) to synthesize cholesterol esters (of cholesterol and fatty acids) results in the formation of low-density lipoprotein (LDL). It is LDL, the formation and structure of which is shown in Fig. 4.28, that provides cholesterol for the plasma membrane of the majority of the body's cells.

Brown and Goldstein found that it was possible to use cultured human fibroblasts as a model system in which to investigate LDL uptake and cholesterol metabolism. These cells have 20 000–50 000 copies of a glycoprotein LDL receptor at the cell surface which recognize the apoprotein B component of LDL and can bind LDL at nanomolar

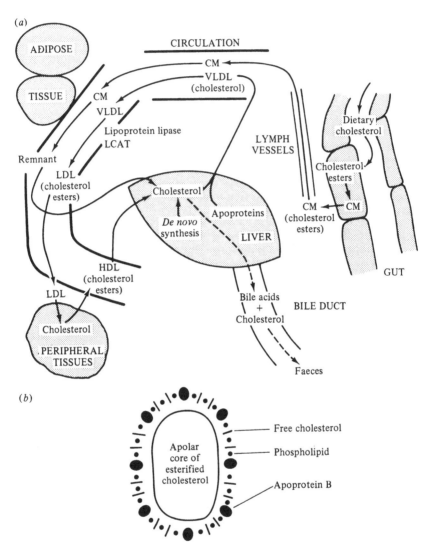

Fig. 4.28. A simplified scheme (*a*) showing the origin of chylomicrons (CM), very low-density lipoprotein (VLDL), low-density lipoprotein (LDL) and high-density lipoprotein (HDL). HDL is synthesized in liver and intestine as a nascent form which collects excess cholesterol from peripheral tissues for return to the liver. HDL is sometimes considered the dustbin of cholesterol metabolism. Exchanges of protein and lipid contents of all the lipoprotein particles can occur and may greatly affect the composition of these in blood. A schematic structure of an LDL particle is shown in (*b*). Such particles have a molecular weight of approx 3×10^6 daltons, are approx 200 Å diameter and have a composition 75 % lipid and 25 % apoprotein B (a glycoprotein).

concentrations. When bound, LDL is internalized by the cells and the process may be followed since cell surface LDL but not internalized LDL can be dissociated with heparin. Internalization is a rapid process, being complete within 10–15 min (at 37 °C), and the internalized LDL is delivered to the lysosomes where cholesterol ester is hydrolysed to release free cholesterol and the protein components are digested to amino acids. The remarkable efficiency of LDL uptake is due to the fact that the receptors are concentrated in coated pits. These contain 80 % of the receptors although they cover only 2 % of the cell surface. The LDL receptors appear to be reutilized (recycled) many times, since it has been shown that LDL uptake can continue for up to 6 h following inhibition of protein synthesis with cycloheximide. Once cholesterol is released in the lysosomes, it can be used for membrane synthesis. It can also regulate intracellular cholesterol concentration by inhibiting cholesterol synthesis (inhibition of 3-hydroxy 3-methylglutaryl coenzyme A reductase), stimulating esterification for storage (stimulation of acylcoenzyme A – cholesterol acyl transferase) and controlling the number of receptors on the cell surface by inhibition of receptor synthesis (Fig. 4.29). The pathway of internalization of LDL appears to be similar in other human cells, including aortic smooth muscle cells, which proliferate in atheros-clerosis and become swollen with droplets of cholesterol esters to pro-duce the characteristic microscopic appearance of 'foam cells'.

The interpretation of the studies on LDL uptake in cultured human fibroblasts was greatly aided by the study of abnormalities found in fibroblasts from patients with genetic disease causing blocks at specific points in the pathway. Thus, information about specific patients led to a

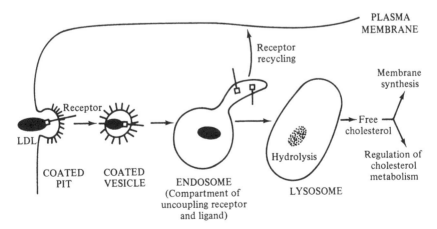

Fig. 4.29. The pathway of endocytosis of LDL in a cultured human fibroblast.

greater understanding of endocytosis and vice versa. The most important mutations were those causing familial hypercholesterolaemia, an auto-somal dominant disease producing high blood levels of cholesterol and premature atherosclerosis. Although the heterozygous form of the disease is common, affecting about 1 in 500 persons of most ethnic groups throughout the world, rare patients (about 1 in 10^6 persons) inherit two copies of the gene and have homozygous familial hypercholesterolaemia which usually results in coronary heart disease before the age of 20. Cultured fibroblasts from such patients did not show the normal respon-ses to LDL (e.g. inhibition of cholesterol synthesis and stimulation of esterification) and were found to possess three types of defect:

1. receptor negative cells, which cannot bind LDL;
2. receptor defective cells, which have only 5–20 % of normal LDL binding capacity;
3. internalization-defective cells, which can bind normal amounts of LDL but cannot internalize bound ligand.

This third defect was particularly interesting, since the normal amounts of LDL receptor on the fibroblast surface were shown to be abnormally distributed: they were not concentrated in the coated pits. The complete amino acid sequence of the normal human LDL receptor has now been deduced by molecular cloning techniques and the abnor-malities in internalization-defective cells from several individuals have been shown to be due to alterations in the cytoplasmic tail of the transmembrane LDL receptor. In one patient, a single base change led to the substitution of a single amino acid in the cytoplasmic tail of the internalization-defective receptor.

Knowledge of LDL receptor endocytosis and the abnormalities of familial hypercholesterolaemia has contributed to ideas about the patho-genesis of atherosclerosis. In particular, it has been suggested that the LDL receptor-mediated endocytosis pathway protects against the disease by allowing cells to satisfy their requirements for cholesterol when the blood concentration is below the threshold range for atherosclerosis. At higher concentrations of cholesterol, receptor-mediated uptake becomes saturated and receptor-independent endocytosis occurs which leads to uncontrolled accumulation of cholesterol ester. Such an event occurring in proliferating arterial smooth muscle cells may cause atheromatous plaque.

Transcytosis

In some cells, the endocytic pathways are modified to allow transcytosis, the movement of plasma membrane components and bound ligands from one side of the cell to another. This process may have considerable physiological significance in the transport of macromolecules across capillary endothelial cells, and across the placenta (for nutrition of the foetus), but also has a function in modifying the composition of some secretions. For example, the major type of antibody found in secretions such as bile, saliva and milk is polymeric immunoglobulin A (pIgA; cf. Chapter 2), which acts as part of the body's immune defence mechanism in these secretions and in milk contributes to the immune defence of the suckling neonate.

In rat liver, the transcytotic pathway by which pIgA moves from blood to bile has been well established using electron microscopic and subcellular fractionation techniques, and identifies several aspects of membrane movement and membrane sorting that remain unresolved. pIgA in the bloodstream binds to a cell surface receptor on the hepatocyte blood sinusoidal plasma membrane which is internalized to the endosomal compartment via coated pits. During this process the pIgA becomes covalently attached to its receptor via a disulphide bond and is not released in the acid interior of the endosome. Glycoproteins that have lost terminal sialic acid (asialoglycoproteins) bind to a different receptor on the blood sinusoidal plasma membrane but can be internalized to the same endosomal compartment. However, while asialoglycoproteins are released from their receptor (which recycles) and are transferred to lysosomes, the pIgA bound to its receptor is segregated into transfer vesicles which pass across the cell and fuse with the bile canalicular plasma membrane. The pIgA–receptor complex is then proteolytically cleaved from the membrane and released into the bile (Fig. 4.30). Strangely, the receptor itself undergoes constitutive transcytosis (i.e. even when pIgA is not bound) and is cleaved and secreted into bile. Indeed, the receptor is called secretory component and has been described as a 'sacrificial receptor'. It is synthesized as a transmembrane protein in the endoplasmic reticulum, transferred to the Golgi apparatus, where glycosylation is completed, and then inserted into the blood sinusoidal membrane. From this membrane it is endocytosed and transferred to the bile canalicular membrane and is thus an example of a plasma membrane protein inserted into one plasma membrane domain before transfer to another.

While the pIgA transcytotic pathway in rat liver is well described (Fig. 4.30), nothing is known of how the receptor becomes associated with coated pits. Unlike the LDL receptor, it is not concentrated in these structures, yet unlike most receptors, it does not require ligand binding

(a)

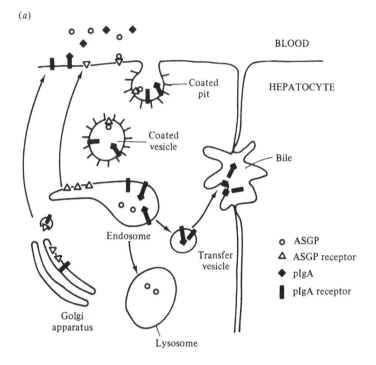

BLOOD

HEPATOCYTE

Coated pit

Coated vesicle

Bile

Endosome

Transfer vesicle

○   ASGP
△   ASGP receptor
◆   pIgA
▮   pIgA receptor

Golgi apparatus

Lysosome

(b)

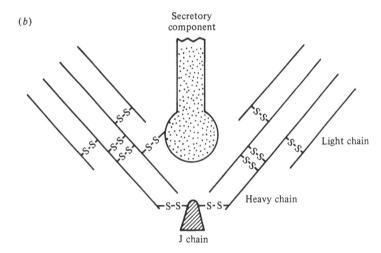

Secretory component

Light chain

Heavy chain

J chain

Fig. 4.30. A comparison of the transcytotic pathway for polymeric IgA (pIgA) in rat liver with the endocytosis of asialoglycoprotein (ASGP) (a). The molecular structure of secretory IgA found in bile is shown in (b) with the pIgA covalently linked to secretory component, a fragment of the pIgA receptor.

for internalization. It is not clear how the pIgA receptor segregates into a transcytotic pathway in the endosome (unlike asialoglycoprotein and its receptor) and little is known of the mechanism of directing the movement of transfer vesicles across the cell (though inhibition by agents such as colchicine that disrupt microtubules suggests that these may be involved).

Transport across the plasma membrane

In so far as the osmotic properties of cells were important in defining the existence of the plasma membrane, it is true to say that transport of water across the plasma membrane was the first observed function of this organelle. Other transport properties were rapidly observed, particularly the fact that lipid-soluble molecules were able to enter cells more easily than charged polar molecules, a fact established by the turn of this century. Despite such early studies and subsequently a vast literature of transport studies and kinetic equations, it is only in the last few years with the isolation, purification and analysis of membrane transport proteins that it has been possible to categorize the permeability properties and mechanisms of the plasma membrane.

The phospholipid bilayer at the cell surface acts as a permeability barrier to most polar molecules, because of its hydrophobic interior. However, small non-polar molecules that can dissolve in the bilayer rapidly diffuse across, as do small, uncharged polar molecules such as water. A molecule such as glucose is already sufficiently large to be impermeant to the bilayer and any charged molecule or ion, however small, is essentially impermeant. The ability of such molecules to cross the plasma membrane is dependent on the presence of integral membrane proteins that themselves cross the phospholipid bilayer. Such membrane transport proteins can function in a variety of ways, either transporting a single molecular species or coupling its transport to that of another species in a co-transport system (Fig. 4.31). Transport may occur by passive diffusion through a channel protein (when diffusion is simply related to the diffusion coefficient of the transported molecule and its concentration gradient), by facilitated diffusion (where binding to the transport, or carrier, protein occurs) or by active transport (where, in addition, energy is required for transport).

Facilitated diffusion resembles an enzyme–substrate reaction and, indeed, the demonstration of Michaelis–Menten kinetics, the temperature coefficient of the process and the effect of specific inhibitors can be used to identify a transport process occurring by this mechanism (Fig. 4.32). The transport proteins involved in such processes appear to be among the largest of intrinsic membrane proteins and are unlikely to

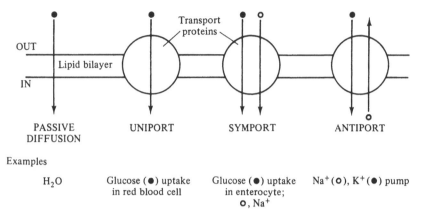

Fig. 4.31. Types of transport process occurring across the plasma membrane.

rotate through the bilayer for thermodynamic reasons. Thus, it is generally thought that a gating mechanism such as that shown in Fig. 4.32 explains the movement of the transported molecules. Only slight modification of such a mechanism is required to account for active transport by coupling the transport process to an energy source, usually ATP. This can allow a cell to accumulate specific substances against considerable concentration gradients (up to 10 000-fold in some cases).

A good example of facilitated diffusion thought to occur by the mechanism shown in Fig. 4.32 is the anion exchange protein of the red blood cell plasma membrane. This protein has been identified as the Band 3 membrane protein on SDS polyacrylamide gels (Fig. 4.22) and accounts for 30 % of the total membrane protein. It is responsible for the exchange of bicarbonate ions formed from tissue CO_2 with chloride ions in the red blood cell cytoplasm. This exchange occurs as the red blood cell passes through tissue capillaries, and the reverse process as the cells pass through the lung. The protein exists as dimers in the membrane and can exchange chloride ions rapidly with a half-time of 50 ms at body temperature. Protein labelling experiments have shown that all copies of the transporter have the same orientation, and it is difficult to conceive of other than a gating mechanism to explain its function.

Recent experiments on two active transport membrane proteins have considerably extended our understanding of how this type of protein functions. The first of these is a bacterial membrane protein, bacteriorhodopsin (not to be confused with animal retina rhodopsin though containing the same retinal chromophore), which forms large purple-

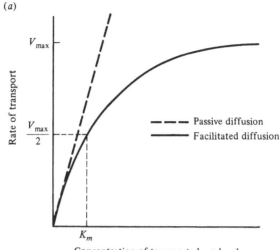

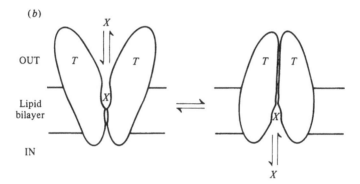

Fig. 4.32. Facilitated diffusion. The kinetics of facilitated diffusion (*a*) are analogous to an enzyme–substrate reaction. It is therefore possible to calculate V_{max} (maximum velocity) of the reaction and K_m (equivalent to the binding constant of the carrier protein for the transported molecule). The gated pore mechanism thought to explain facilitated diffusion is shown in (*b*). The transported molecule X forms a transient complex with carrier protein T. This results in a conformational change in T allowing X to pass across the membrane. Net transport of X down a concentration gradient of X will occur.

coloured patches on the surface of *Halobacterium halobium* when the organism is grown in the absence of oxygen. Its function is to generate an electrochemical gradient of protons across the membrane that can in turn be used to generate ATP. Because bacteriorhodopsin is present in an almost perfect crystalline array in the bilayer, it has been possible to use

X-ray analysis to show its structure. It has been found to consist of trimers of the protein each having seven α-helices spanning the membrane. The energy for the active transport carried out by bacteriorhodopsin is derived from light, so that when a single photon of light activates the chromophore retinal, it causes a conformational change in the protein resulting in the transfer of two H^+ from the inside to the outside of the cell. In this example, it seems that rather than the conformational change occurring between two subunits of a multimeric protein, it occurs between transmembrane segments of a single polypeptide chain.

The $Na^+ K^+$ pump found in the plasma membrane of almost all animal cells is a more representative example of an active transporter since it uses ATP as its energy source. Indeed, its ATP requirement can account for more than 30 % (70 % in nerve cells) of the total energy utilization of the cell. $Na^+ K^+$ ATPase has been purified and shown to be a tetramer of two transmembrane catalytic subunits (Mr 100 000) and two molecules of an associated glycoprotein (Mr 45 000). The catalytic subunit binds K^+ on the external surface and Na^+ and ATP on the cytoplasmic surface, ATP causing a reversible phosphorylation of the protein. For every molecule of ATP hydrolysed, three Na^+ are pumped out and two K^+ pumped in (Fig. 4.33). This helps to maintain the ionic composition of the cytoplasm and the electrochemical potential of the plasma membrane, though the latter is also affected by a K^+ leak channel which allows K^+ to leak out of the cell down its concentration gradient, and the permeability of the plasma membrane to Cl^-. The $Na^+ K^+$ ATPase is also important in

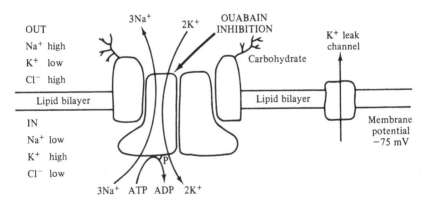

Fig. 4.33. The structure and function of $Na^+ K^+$ ATPase. The catalytic subunit has a large cytoplasmic domain containing several loops of the polypeptide chain. It should be noted that the separate potassium leak channel is thought to be responsible for 80 % of the membrane potential and that this membrane potential prevents Cl^- influx down a concentration gradient into the cell.

controlling the osmotic properties of the cell. If the inhibitor ouabain is added to isolated cells, it binds to the K^+ binding site on the extracellular face of the carrier protein, inhibiting its function and causing the cells to swell and burst since they are no longer able to regulate their intracellular ionic composition.

Some active transport systems can be driven by the energy stored in ion gradients rather than ATP. These include some sugar and amino acid transport systems. A good example is the symport of glucose and Na^+ into intestinal epithelial cells from the gut lumen (Fig. 4.31), though in this case, the system is complicated by the need for transcellular transport of glucose to the blood, which is achieved by a different glucose carrier protein capable of facilitated diffusion in the basolateral plasma membrane.

Disease and the plasma membrane

Abnormalities of many aspects of membrane structure and function have been found and associated with specific disease conditions. At a broad structural level, alterations in membrane lipid composition and fluidity have been recorded in red blood cell and other plasma membranes in many diseases. In the case of atherosclerosis, it is likely that high serum cholesterol concentration reflected in plasma membrane composition leads to increased membrane rigidity in the cells of arterial walls and alterations of cellular function. It is very much less clear whether the increase in membrane fluidity observed in other diseases including muscular dystrophy and cystic fibrosis or in many transformed cells is related to the primary lesion of the disease.

Structural abnormalities at the level of interaction of the cytoskeleton with the plasma membrane have also been reported, the best described being red blood cell disorders. Hereditary elliptocytosis is a group of disorders, some of which cause severe haemolytic anaemia. The frequency of all forms is approximately 1:4000. In about half of the patients, the disorder has been shown to be an abnormality of the cytoskeleton, either in spectrin or in one of the other proteins linking the cytoskeleton to the lipid bilayer. These abnormalities result in the red blood cells having an unusual elliptical shape. Perhaps the best known of all diseases explained at a molecular level is sickle cell anaemia, caused by a single amino acid substitution in the haemoglobin β chain. The characteristic sickling of the red blood cells in this condition is apparently due to fibrous bundles of polymerized abnormal haemoglobin molecules protruding through the cytoskeleton and causing local uncoupling of the cytoskeleton and the lipid bilayer.

Hereditary disorders of specific membrane functions have also been described. These include abnormalities of receptor function (e.g. LDL receptor, see above), transport systems and membrane enzymes. Among the best documented are genetic defects in transport systems in the brush border of intestinal cells. In these cells, abnormalities of sugar, amino acid and phosphate transport are known, each resulting in specific nutritional deficiency diseases. Even more intriguing is the possibility that an abnormality of ion transport across the plasma membrane may be responsible for the development of cystic fibrosis, a major hereditary disease. It has been suggested that the primary defect in this disease is an abnormality in a chloride channel, though it is likely that the DNA probe approaches described in Chapter 7 will need to get closer to the major gene involved before this hypothesis can be fully tested.

In addition to hereditary defects, it is possible for membranes to be damaged during the development of a disease or as a primary event causing disease. The ability of coated viruses to be endocytosed and then

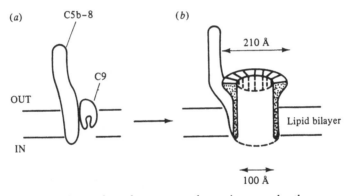

Fig. 4.34. Formation of a transmembrane ion pore by the complement membrane attack complex. When antibodies bind to a cell surface, they may activate a series of about 20 serum proteins known as complement components. These are activated in a strict order and can form a membrane attack complex composed of the five terminal components which are C5b, C6, C7, C8 and C9. It is thought that a C5b-8 complex is formed as an integral membrane protein. This acts as a receptor for soluble monomeric C9 (a glycoprotein of $Mr\,70\,000$) which binds and disturbs the lipid bilayer (*a*). Monomeric C9 is triggered to polymerize, forming a transmembrane poly C9 complex which is a stable lesion in the membrane (*b*) and can be seen by electron microscopy. C9 insertion into the target membrane allows the passage of ions and small molecules, leading to metabolic damage and, as a result of alteration in osmotic pressure, to cell bursting and death.

fuse with the endosome membrane allowing the infectious nucleic acid into the target cell cytoplasm has already been described. Proteins may also damage cell membranes, examples being found among toxins, venoms and components of the immune system. Cholera toxin binds to a specific membrane lipid, inserts into the target membrane and activates the production of cyclic AMP (see Chapter 6). Other bacterial toxins and mellitin, the major protein component of bee venom, are capable of forming transmembrane ion channels in target cells, damaging and potentially killing the cell. The terminal protein component (C9) of the complement system found in mammalian serum can similarly become a transmembrane protein (Fig. 4.34). The complement cascade is triggered by antibodies binding to invading bacteria and C9 insertion into the bacterial membrane helps prevent infection. When the immune system makes the error of reacting with normal host tissues, autoimmune diseases may occur. These can be the direct result of an interaction of an antibody or immune cells with a specific membrane protein. In some cases, the complement cascade can be triggered, resulting in C9 insertion into the membrane, causing or contributing to cell damage.

Damage to the plasma membrane in disease can provide helpful diagnostic information. Not only may intracellular proteins be specifically released by the damaged cells in many diseases (see Chapter 3), but, for example in the case of obstructive jaundice, some liver plasma membrane enzymes are also released into the circulation. The integral membrane ectoenzymes 5'-nucleotidase and alkaline phosphatase are naturally shed from the surface of hepatocytes into the bile and, if the bile duct is blocked, they enter the bloodstream. Despite its diagnostic usefulness, the mechanism of this process is not clear.

5

Hormone action

Intercellular communication

The integration and coordination of cellular activities in a multicellular organism requires communication between cells. At the simplest level, intercellular communication may result directly from cell contact. In mammalian tissues, even direct cell contact may be modified by the appearance of specific plasma membrane modifications. Thus, gap junctions (Chapter 4) allow the passage of small molecules and ions between adjacent epithelial cells. More complex modifications may also occur as between nerve cells and the target cells they influence, where formation of a synapse allows the nerve cell to release a chemical messenger (neurotransmitter) on to the target cell without affecting surrounding cells. Cells can also communicate without contact if chemicals released by one group of cells affect another group. This communication may be local if the chemicals are rapidly taken up or destroyed, or over longer distances if they are carried in the body fluids. The actions of such chemicals may be acute, in which case the chemicals tend to be regarded as hormones, or long-term, when the chemicals are often considered as growth factors. This distinction is somewhat artificial as hormones with both acute and chronic effects have been described (see below). In all cases of cell–cell interaction, the target cells respond to the extracellular signal by alterations in their intracellular metabolism. The underlying biochemical mechanisms of such intracellular alterations appear to be relatively few, yet they control the growth, development, organization and diversity of all mammalian tissues.

Hormones as chemical messengers

The classical definition of a hormone is that proposed by Starling in 1905 after his discovery that the peptide hormone gastrin controls acidification in the stomach. Starling defined hormones as substances 'produced for effecting the correlation of organs within the body, carried from the organ where they are produced to the organ which they affect by means of the blood stream'.

While Starling's definition of a hormone remains appropriate, it fails to emphasize the similarities existing between a variety of extracellular regulators, and an alternative definition proposed by Huxley (1935) may be preferred. This states that 'hormones are information-transferring molecules, the essential function of which is to transfer information from one set of cells to another for the good of the population as a whole'. Although Starling's definition clearly distinguishes hormones and neuro-transmitters, since only the former are delivered by the bloodstream, Huxley's definition of a hormone allows the inclusion of many locally produced chemicals that are capable of modifying hormone action and/or have their own specific effects. Such chemicals include prostaglandins, peptides and metabolites such as adenosine, all of which may act as 'local hormones'.

Even within Starling's definition, the number of hormones is large, though chemically they may be subdivided into steroids, amino acid derivatives (thyroid hormones and catecholamines) and peptide hormones (Table 5.1). The latter can range in size from three amino acids (thyrotropin releasing hormone) to hundreds of amino acids and may have subunits. Despite the variety of hormones and the great diversity of tissues and physiological functions affected, hormones have a number of common features:

1. Cells producing hormones are known as endocrine cells and secretion can be triggered by specific metabolites or other hormones.
2. Hormones can act on target tissues at very low concentrations, often circulating in the bloodstream at concentrations of less than 10^{-9} M.
3. Hormones show specificity in causing responses in different tissues and different cell types. They activate or inhibit selected metabolic processes, some hormones triggering multiple events in a single cell type and others triggering cell specific events in different cell types.
4. Hormones interact in their effects on metabolic processes. Both synergistic and antagonistic interactions are known.
5. Hormones may cause acute or chronic effects:
   (a) acute effects – occur within seconds or minutes
                       – magnitude of response is proportional to hormone concentration
                       – usually of short duration

Table 5.1 *The variety of hormones.*
Representative examples from each class of hormone are shown.

Hormone	Site of origin (endocrine gland)	Major target tissue(s)	Major effects	Structure
Steroids				
Oestradiol	Ovary (follicle)	Breast, uterus, vagina	Maintenance of secondary female sex characteristics, maturation and normal cyclic function of accessory sex organs, development of ducts in mammary gland.	
Amino acid derivatives				
3,5,3 triiodothyronine (T3)	Thyroid	Many cells	Increases metabolic activity	
Adrenalin	Adrenal medulla	Heart, muscle, liver	'Fight, fright and flight' hormone increases blood pressure and heart rate, increases glycogen breakdown in liver and muscle	

Peptide hormones

Thyrotropin releasing factor (TRF)	Hypothalamus	Anterior pituitary	Secretion of thyroid stimulating hormone	3 amino acids
Glucagon	Pancreatic α cells	Adipose tissue, liver	Stimulation of lipolysis and glycogen breakdown	29 amino acids
Insulin	Pancreatic β cells	Muscle, adipose tissue, liver	See Table 5.3	2 peptide chains, A chain 21 amino acids, B chain 30 amino acids joined by disulphide bonds
Follicle stimulating hormone (FSH)	Anterior pituitary	Ovary	Follicle development, secretion of progesterone	Glycoprotein, α chain 92 amino acids, β chain 108 amino acids
Thyrotropin (TSH)	Anterior pituitary	Thyroid	Stimulation of thyroid hormone secretion	Glycoprotein, α chain 92 amino acids (in common with FSH), β chain 113 amino acids

Local hormones

Prostaglandin E$_2$	Many cells	Smooth muscle	Contraction	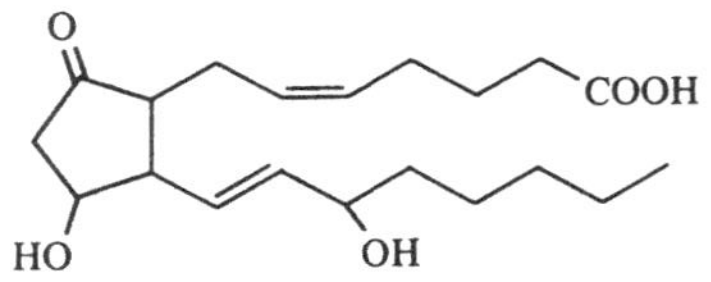
Adenosine	Many cells	Many cells	Vasodilation, modulation of adenylate cyclase	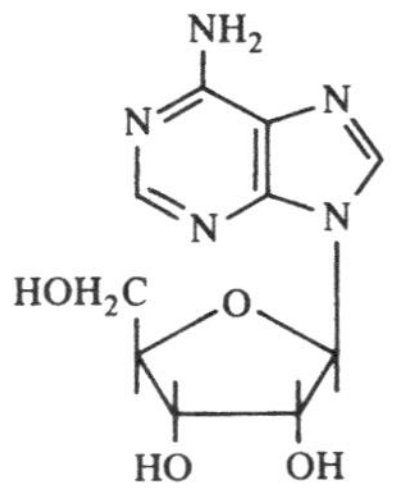

(*b*) chronic effects – take hours to become apparent
– persist for long periods
– independent of short-term fluctuations in hormone concentration.

Catecholamines and many polypeptide hormones cause acute effects, whereas chronic effects are caused by steroids, thyroid hormones and some polypeptide hormones. Polypeptide hormones such as glucagon and insulin have both acute and long-term actions.

6. Specific degradation processes exist, often rapidly acting, to inactivate the hormone.

In spite of the low circulating concentrations of hormones ($<10^{-9}$M), they can trigger changes in cell metabolites in the 10^{-3}M range. Thus, cells can be capable of a 10^6-fold amplification of response to hormone action. Although some hormones, specifically steroid hormones and thyroid hormones, are lipid-soluble and enter cells across the plasma membrane phospholipid bilayer, the majority are water-soluble and cannot cross the plasma membrane. Instead, they exert their effects by binding to high-affinity cell surface receptors, many of which have been identified as transmembrane glycoproteins. It is now clear that the acute effects of water-soluble hormones occur as the result of specific membrane events caused by receptor binding. In addition, it is apparent that the number of available biochemical mechanisms by which hormones may activate cell processes is very limited relative to the number and variety of hormones and physiological functions affected. The best understood mechanism, activation of adenylate cyclase which catalyses cyclic AMP production, is described below, along with well documented examples of metabolic pathways affected by a rise in the intracellular concentration of cyclic AMP. This leads to a discussion of hormones with a more complex mechanism of action such as insulin and steroid hormones and finally some consideration of hormone interactions which control physiological function.

Experimental investigation of hormone action and hormone binding

In order to identify the molecular and biochemical events involved in altering physiological function, it may be necessary and is certainly possible to investigate hormone action at the level of the whole animal, perfused organ, tissue culture, isolated cells, cell homogenate, subcellular fraction or even purified enzyme complex. It is often easy to demonstrate that a given hormone has effects on cells or tissues but much more difficult to show that these are of physiological importance. In this context, it is important to compare the dose response curve and time course of hormone action *in vitro* with a knowledge of the changes in hormone concentration found *in vivo*. There is little point in showing that a 10^{-3}M solution of a hormone will activate a cellular process if the normal concentration to which the cell is exposed *in vivo* is 10^{-9}M. Effects observed *in vitro* at 10^{-3}M may be non-specific effects of the hormone or due to the presence of an impurity, i.e. another hormone present at much lower concentration.

The interaction of a hormone with its cellular receptor can be described and quantified using techniques originally applied to the binding of drugs to their receptors. It is usually necessary to use a radioactively labelled hormone for such studies and it is imperative that the labelled molecules remain biologically active and that the measured kinetics, affinity and structural specificity of hormone binding to receptors are consistent with the corresponding characteristics of the hormone-stimulated biological responses. In general, the number and affinity of cell surface hormone receptors may be estimated by Scatchard analysis (Fig 5.1). Receptors are usually found to be of high affinity, very specific and present at highest concentration in recognized target tissues for the hormone. Molecules with very similar structure may antagonize binding (Fig. 5.2) and can be used to help elucidate structural features of the hormone necessary for binding and action. Some hormones, e.g. catecholamines, are capable of interacting with more than one kind of receptor which are the products of distinct genes and mediate different biochemical events. Other hormones, e.g. insulin, probably have only one kind of receptor which is very similar in different tissues, though tissue variation based on post-translational modification, e.g. glycosylation, may subtly affect the hormone interaction.

Most acute-acting hormones have between 10^4 and 10^5 high-affinity receptors on the surface of a single target cell. This number of receptor molecules per cell implies that they are minor membrane components and they have proven difficult to purify, often requiring 50 000-fold purification relative to total homogenate protein. The protein nature of receptors was initially shown by the action of proteolytic enzymes to destroy the binding sites with simultaneous loss of the hormonal sensitivity of cell and membrane preparations. Despite the knowledge of the primary structure of some receptors, e.g. insulin receptor (see below), achieved through using recombinant DNA techniques (see Chapter 7), the majority of receptors are still largely uncharacterized at the level of

It is assumed that the hormone (H) binds to its receptor (R) in a reversible manner and that there is no interaction between receptor molecules or between hormone molecules.

$$R + H \rightleftharpoons RH$$

Then by the Law of Mass Action

the equilibrium constant $K = \dfrac{[RH]}{[R][H]}$

If B represents the concentration of bound hormone and F the remaining concentration of free hormone, this simplifies to

$$K = \frac{B}{([R] - B)F} \quad \text{or} \quad \frac{B}{F} = K([R] - B)$$

Thus, an experiment measuring bound and free hormone at different hormone concentrations will yield the graphical representation (Scatchard plot):

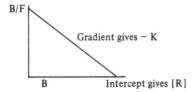

Thus, the number of binding sites [R] and the equilibrium constant K can be determined. This form of K is usually referred to as the association constant (K_A in units of litres per mole) or dissociation constant (K_D in units of moles per litre). K is larger when the binding between R and H is stronger.

In some cases, it is known that receptors interact giving non-linear Scatchard plots. There is evidence that some hormones become covalently linked to receptor after initial reversible binding. Such events can complicate binding studies.

Fig. 5.1. Scatchard analysis of hormone binding to a tissue receptor.

structure or primary sequence. Experiments using immobilized but still biologically active hormones covalently coupled to large inert beads suggested that hormone binding to the cell surface was sufficient to generate hormonal responses and internalization of hormones was unnecessary. Such results led to the question of how hormone binding at the cell surface can cause intracellular metabolic changes. It is now clear that these changes are the result of membrane events leading to the production of specific second messengers and/or the movement of ions.

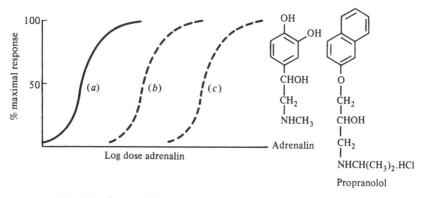

Fig. 5.2. Competitive inhibition of hormone action by an analogue. Dose response curves reach the same maximum but are shifted to the right as the concentration of competitive antagonist is increased. The β blocker propranolol with a modified catechol group is a competitive antagonist of adrenalin action. (*a*) adrenalin alone, (*b*) and (*c*) adrenalin plus propranolol. The propranolol concentration in (*c*) is greater than in (*b*).

Cyclic AMP as a hormone second messenger

Our understanding of the mechanism of hormone action was revolution-ized in the 1950s as a result of the experiments of Sutherland and his colleagues, who were investigating the breakdown of liver glycogen in response to adrenalin. They were able to demonstrate the effect of the hormone on glycogen breakdown in dog liver slices, the first well documented hormone sensitive *in vitro* system. More importantly, they showed that adrenalin caused liver membrane fractions to produce a heat-stable factor that caused the activation of phosphorylase, the enzyme stimulating glycogen breakdown. The heat-stable factor was identified as cyclic AMP, produced by the action of the intrinsic plasma membrane enzyme adenylate cyclase on ATP (Fig. 5.3). Cyclic AMP was found to be ubiquitous in eukaryotic tissues and also to play an important role in metabolic control in some prokaryotes. In mammalian cells, it was shown to be the second messenger for a wide variety of hormones, including, for example, glucagon, thyrotropin, adrenocorticotrophic hormone as well as catecholamines. The following criteria were adopted to demonstrate that a hormone acted via cyclic AMP:

1. The hormone affects cyclic AMP concentration in the target tissue with an appropriate dose response curve and time course.
2. Inhibitors of cyclic AMP phosphodiesterase potentiate the action of the hormone by preventing cyclic AMP degradation.

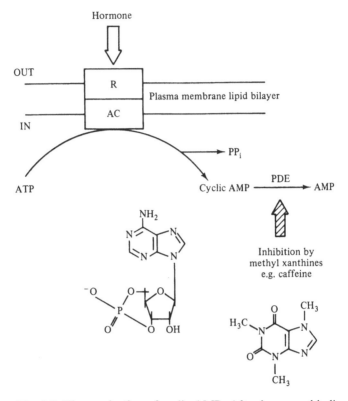

Fig. 5.3. The production of cyclic AMP. After hormone binding to the cell surface receptor (R), adenylate cyclase (AC) is activated. The enzyme has a K_m for MgATP of approximately $0.15\,\mu M$ and can therefore function at maximal velocity at the intracellular ATP concentration (approximately 5 mM). The cyclic AMP produced acts as a hormone second messenger within the cell. Pyrophosphate (PP_i) is rapidly cleaved to inorganic phosphate, helping to drive the adenylate cyclase reaction. Methyl xanthines (e.g. caffeine) can inhibit cyclic AMP degradation by phosphodiesterase (PDE) and potentiate hormone action.

3. The lipid-soluble cyclic AMP analogue dibutyryl cyclic AMP mimics the action of the hormone (since it can cross the plasma membrane).

4. The hormone can stimulate adenylate cyclase in homogenate and membrane preparations.

Analogous criteria should be applied in considering other second messengers, though this is not always easy to do.

The coupling of hormone-receptor binding to activation of adenylate cyclase

Early experiments on adenylate cyclase showed that its active site was on the cytoplasmic side of the plasma membrane, i.e. the opposite side to the receptor binding site for hormone. This was most clearly demonstrated in experiments on avian red blood cells which have an adrenalin stimulated adenylate cyclase (N.B. mammalian red blood cells do not have the enzyme). These experiments showed that the cells could not form cyclic AMP from extracellular ATP, that cyclic AMP was formed intracellularly in response to adrenalin and that adenylate cyclase could not be destroyed by trypsin in intact cells but was rapidly destroyed in lysed cells. Thus, coupling of hormone binding to activation of adenylate cyclase is a transmembrane event.

A second observation was that different hormones could activate adenylate cyclase in the same cell preparation, suggesting the possibility of adenylate cyclase isoenzymes. In fact, there is only one type of adenylate cyclase which may be coupled to different receptors. Thus, submaximal amounts of different hormones may give an additive response, but when the enzyme is fully activated by one hormone, a different hormone cannot provide further stimulation. Specificity of hormone action occurring via an intracellular rise in cyclic AMP concentration is therefore due to the presence or absence of specific cell surface receptors and not subsequent events.

A third, possibly surprising, observation was that for many hormones the maximum biological effect could be obtained when only a fraction of hormone receptors was occupied. It was shown, for example, that in pig kidney plasma membranes stimulated by anti-diuretic hormone, 80 % of maximal adenylate cyclase activity could be achieved with only 10 % of maximal hormone binding. Similar observations about the relationship of biological response to receptor occupancy have been made even for hormones that do not act by stimulating adenylate cyclase. Such experiments have given rise to the concept of spare receptors, though an apparent excess of receptors may be necessary to allow occupation of a sufficient number at low hormone concentration, given the available binding affinity.

Many schemes were proposed to explain the molecular linking of hormone receptor to adenylate cyclase, usually involving the existence of multi-protein complexes in the membrane. However, cell fusion experiments showed that permanent complexes did not exist and that a hormone receptor supplied by the plasma membrane of one cell could activate adenylate cyclase supplied by another (Fig. 5.4). Similar experi-

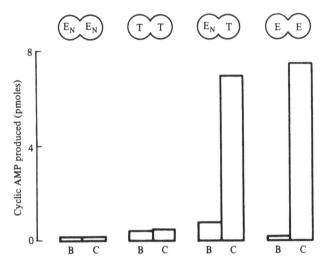

Fig. 5.4. A typical cell fusion experiment to show the interaction of the hormone receptor of one cell with the adenylate cyclase of another. Turkey red blood cells (E) possess a catecholamine activated adenylate cyclase. The enzyme can be inactivated without affecting catecholamine binding by treating the cells with N-ethylmaleimide (E_N) or heat. Cultured rat adrenal tumour cells (T) have an adenylate cyclase which can be activated by adrenocorticotrophic hormone (ACTH). Cells may be fused in the combinations indicated using Sendai virus. The presence in the fused cells of an active adenylate cyclase (or its absence in E_N cells) can be shown by challenging with fluoride which results in persistent activation of the enzyme through an effect on the G protein. Fused E_N–T cells show a catecholamine stimulated adenylate cyclase. B, basal activity; C, plus $50\,\mu$M catecholamine. Data redrawn from Schramm, M. *et al.* (1977). *Nature*, **268**, 310–13.

ments were carried out using membranes and even partially purified receptors, showing that receptors interact with adenylate cyclase as a result of lateral mobility within the fluid phospholipid bilayer.

The coupling of hormone receptor to activation of adenylate cyclase is mediated by an additional membrane protein, a regulatory unit known as G protein or N protein that is also fluid within the plane of the bilayer. Initial evidence for the presence of G protein came from studies on liver plasma membrane adenylate cyclase in which it was shown that GTP (guanosine triphosphate) potentiated the action of glucagon to stimulate the enzyme. In fact, GTP itself activated adenylate cyclase, an effect more apparent when a non-hydrolysable analogue of GTP was used (Fig. 5.5). Further evidence for the G protein was obtained from cultured

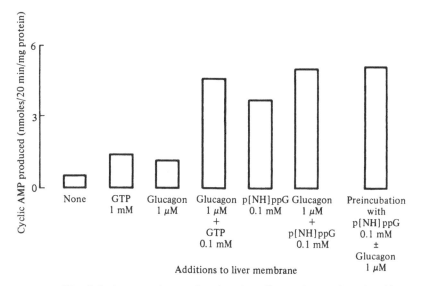

Fig. 5.5. An experiment showing the effects of guanyl nucleotides on rat liver plasma membrane adenylate cyclase. Note that preincubation (10 min, 30 °C) with the non-hydrolysable analogue of GTP p[NH]ppG (guanosine–5'–[βγ imido] triphosphate) causes maximal stimulation of the enzyme. Data from Martin, B. R. *et al.* (1979). *Biochemical Journal*, **184**, 253–60.

mutant cell lines with normal amounts of hormone receptor and adenylate cyclase but no coupling between them. Fusing membrane preparations from these cells with detergent solubilized membrane fractions from normal cells reconstituted hormonal activation of adenylate cyclase. The G protein, as well as activating adenylate cyclase, is itself a GTPase converting GTP to GDP and appears only to activate adenylate cyclase when GTP is bound, not GDP. The G protein can be activated irreversibly by non-hydrolysable derivatives of GTP and also by the action of cholera toxin which penetrates the membrane and ADP ribosylates the G protein preventing GTP hydrolysis (Fig. 5.6).

The overall mechanism by which hormones activate adenylate cyclase is thought to be that shown in Fig. 5.7. The hormone binds to its receptor causing a conformational change and allowing the receptor to interact with inactive G protein. This interaction in turn enables GTP to displace GDP, altering the conformation of the G protein so that it may bind and activate the catalytic subunit of adenylate cyclase. When the G protein hydrolyses the GTP, it returns to its original conformation, causing the adenylate cyclase to dissociate and become inactive. A noticeable feature

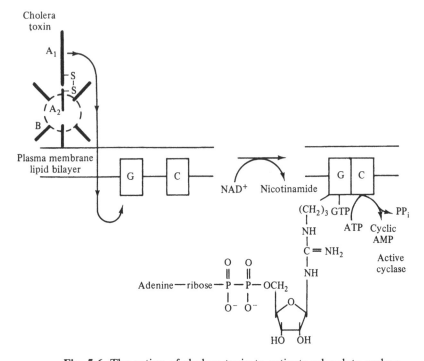

Fig. 5.6. The action of cholera toxin to activate adenylate cyclase. Cholera toxin consists of an A_1 peptide linked by a disulphide bond to an A_2 peptide non-covalently linked to five B peptides. It binds through the B peptides to a specific ganglioside on the cell surface, allowing entry across the membrane of the A_1 subunit. This catalyses transfer of ADP-ribose from intracellular NAD^+ to an arginine side-chain of the G protein (α subunit of G_s) altering it so that it can no longer hydrolyse bound GTP. In intestinal epithelial cells, the resultant chronic activation of adenylate cyclase and prolonged elevation of cyclic AMP concentration causes a large efflux of Na^+ and water into the gut to give the classic symptoms of severe diarrhoea associated with cholera.

of this scheme is that the envisaged collision-coupling allows one hormone receptor complex to activate several catalytic units before the latter returns to the inactive state and the hormone dissociates from the receptor. Thus, the hormone may cause a large stimulation when only a proportion of receptors are occupied. Also, if one type of receptor is regulated on the cell surface, this will not affect the ability of other types of hormone receptor complex to maximally activate the adenylate cyclase.

While the above scheme provides a simple overview of hormone

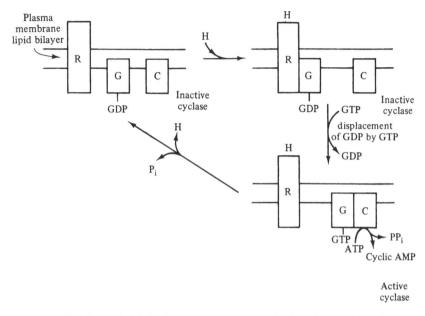

Fig. 5.7. Model for hormone activation of adenylate cyclase. R, hormone receptor; H, hormone; G, G protein; C, catalytic subunit of adenylate cyclase.

activation of adenylate cyclase, the reality is more complex. In particular, it is now clear that more than one type of G protein can regulate adenylate cyclase. Whereas hormones activating adenylate cyclase act via a stimulatory G protein now known as G_s, those inhibiting the enzyme (e.g. the peptide hormone vasopressin) act via an inhibitory G protein called G_i (Fig. 5.8). This was identified by studying the effects of pertussis toxin which abolishes hormonal inhibition of adenylate cyclase by catalysing ADP-ribosylation of G_i. The G protein affected by pertussis toxin (G_i) was shown to be distinct from that affected by cholera toxin (G_s) and in most tissues is present in approximately 30-fold excess over G_s. Both G proteins are made up of three polypeptide subunits denoted α, β and γ, the latter two being common to G_s and G_i. In the activation of adenylate cyclase by stimulatory hormones, it appears that, after binding GTP, G_s dissociates, the α subunit with bound GTP activating the catalytic unit of the enzyme. When inhibitory hormones bind to their receptors, G_i is activated by binding GTP. This also results in dissociation of the α subunit with GTP bound from $\beta\gamma$. Since G_i is present in the membrane in larger amounts than G_s, the excess $\beta\gamma$ complex released acts to reverse and inhibit G_s activation. Thus, direct effects of G proteins on the catalytic

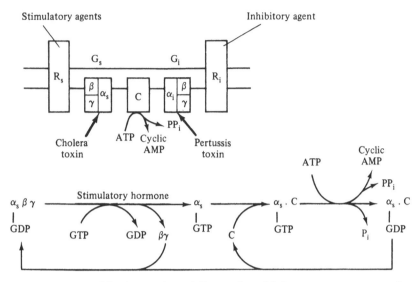

Fig. 5.8. The interaction of G proteins with hormone receptors and the catalytic subunit of adenylate cyclase. R_s and R_i, stimulatory and inhibitory receptors, respectively; C, catalytic subunit. A reaction sequence for activation of the catalytic subunit is shown below the block diagram. Human platelet membranes have a well studied dual regulated adenylate cyclase in which activation has been shown to occur as in the diagram via, for example, glucagon, vasopressin and adrenocorticotrophic hormone, and inhibition via, for example, angiotensin II and opioids.

subunit of adenylate cyclase may be mediated entirely by the G_s α subunit.

Signal amplification and the control of metabolic pathways

Hormones at 10^{-9}M concentration in the circulation can cause 10^{-3}M changes in the concentration of cell metabolites, yet if they stimulate adenylate cyclase, the concentration of cyclic AMP is unlikely to rise above 10^{-6}M. Therefore, despite the amplification occurring as a result of collision-coupling mechanisms in the plasma membrane and the enzymic activity of adenylate cyclase, further signal amplification is required in the cell. This is achieved by the activation of cyclic AMP-dependent protein kinase which can phosphorylate and thereby regulate key enzymes in metabolic pathways (Fig. 5.9). It appears that activation of protein kinase is the only mechanism by which the effects of cyclic AMP are expressed.

Cyclic AMP-dependent protein kinase consists of regulatory and catalytic subunits and is a tetramer in the inactive state. Binding of cyclic

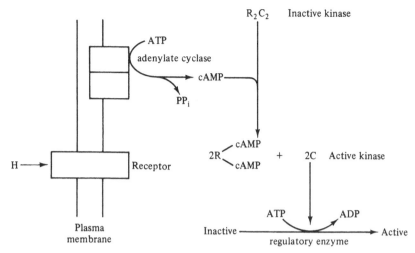

Fig. 5.9. The activation and mode of action of cyclic AMP (cAMP) dependent protein kinase to stimulate a metabolic pathway. The amino acid sequence at the phosphorylation site of the regulatory enzyme is often –Arg–Arg–X–Ser(P)–, where X may be one of several different amino acids and (P) represents the covalently linked phosphate.

AMP to the regulatory subunits releases free active catalytic subunits which can use ATP as a substrate to phosphorylate covalently a wide variety of intracellular proteins. Phosphorylation occurs on serine residues, close to a pair of basic amino acids, in the target proteins. In a number of cases, it has been shown that phosphorylation of a target enzyme *in vivo* or *in vitro* can account for the hormone-stimulated activation of the enzyme. While cyclic AMP-dependent protein kinase is promiscuous in phosphorylating many proteins, the particular metabolic response characteristic of cyclic AMP mediation in a particular tissue will depend on which protein substrates are present and available in that tissue. Although many enzymes are activated by cyclic AMP-dependent phosphorylation, others are inactivated, so that the overall biochemical result of cyclic AMP-dependent hormone action is the switching off of some metabolic pathways as well as the switching on of others. An activated phosphorylated enzyme, by catalysing the transformation of many molecules of substrate to product, completes the amplification of the hormone response.

Control of metabolic pathways by hormones is not achieved by affecting many enzymes in the pathways, but simply those key enzymes involved in rate-limiting steps. In any one pathway, under steady-state conditions, the rate of net substrate utilization equals the rate of net product output. Thus, the net flux through each enzyme-catalysed step is the same and cannot be greater than the maximum possible flux catalysed by the enzyme with lowest activity. A variety of theoretical and experimental approaches can be used to determine the rate-limiting steps of metabolic pathways (Fig. 5.10) and the identification of these steps is

1   The rate-limiting step is often near the start of a metabolic pathway. It is thought that this may be because it is energetically wasteful to build up unwanted intermediates.

2   Measurement of enzyme activities *in vitro* may indicate that some enzymes are present in great excess relative to the activity required for the maintenance of the metabolic pathway. These enzymes are unlikely to be regulatory. Enzymes with low activity may be regulators. Enzymes affected by products of the pathway may be regulators.

3   Rate-limiting steps are sometimes called 'non-equilibrium' steps since the ratio of product to substrate steady-state concentrations is much less than the equilibrium ratio. The mass action ratios of steady-state intracellular concentrations for a reaction may be measured and compared with the equilibrium constant for the reaction.

4   In a pathway $A \to B \to C \to D \to E \to F \to G$, if $B \to C$ is rate-limiting, then addition of C, D, E and F should rapidly produce G, whereas addition of A and B will produce G more slowly.

5   If conditions are rapidly changed such that a metabolic pathway is switched off, then the concentration of intermediates before the rate-limiting step will rise and those after it will fall. This is known as a *cross-over* and may be achieved by adding an inhibitor or altering environmental conditions, e.g. the effect of 5 min aerobic perfusion after anoxia on glycolytic intermediates in perfused rat heart shows that phosphofructokinase (converting fructose 6-phosphate to fructose 1, 6-bisphosphate) is a rate-limiting enzyme in glycolysis.

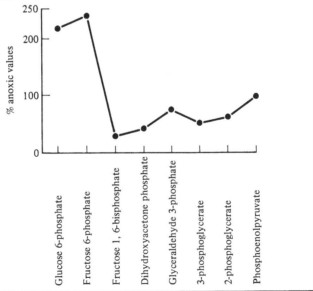

Fig. 5.10. Identification of a rate-limiting step in a metabolic pathway.

Table 5.2 *Chemical composition and energy stores of a 65-kg man*

	Total composition (kg)	% whole weight	Available for oxidation kg	kcal	Daily utilization* during fasting (g)
Protein	11.5	17.7	2.4	9 600	60
Carbohydrate	0.5	0.8	0.15	600	All used in 1–2 days
Lipid	9.0	13.8	6.5	58 500	150
Water	40.0	61.6			
Minerals	4.0	6.1			

*Assuming 1600 kcal per day utilized in the fasting state.
Table from Campbell, A. K. & Hales, C. N. (1976). *Cell Med. Sci.*, **4**, 105–51. It should be noted that, during starvation, while the body's energy supply is derived mostly from fatty acids, the brain retains an absolute requirement for glucose. This is supplied by gluconeogenesis in the liver using amino acids from skeletal muscle proteolysis as substrate.

important in indicating potential sites of hormonal control. It is, of course, possible that a metabolic pathway may contain more than one regulatory enzyme, since the rate limiting step may change under different metabolic conditions.

Despite our general understanding of the mechanism by which some hormones affect metabolic pathways via an activation of adenylate cyclase, a knowledge of such molecular and biochemical events may not totally explain the physiological response of the target tissue. Thus, the half-maximal stimulation of the overall physiological response of a tissue by a hormone may reflect the maximal stimulation of half the cells rather than a submaximal response by all. Such considerations may not be trivial *in vivo*, where cells are likely to be at different developmental stages and in different microenvironments.

The control of lipolysis
Adipose tissue triacyl glycerol is the major energy store in Man (Table 5.2) and its breakdown (i.e. lipolysis) to provide free fatty acids and glycerol for use by muscle and other tissues is regulated by catecholamines, glucagon and insulin. The importance of fatty acids as metabolic fuels under normal physiological conditions is shown by the respiratory quotient (RQ) of Man after an overnight fast which is about 0.75. It has been estimated that 50 % of total respiration at this time is due to fatty acid oxidation, compared to 18 % due to glucose oxidation, most of which occurs in the brain. During starvation, blood glucose concentra-

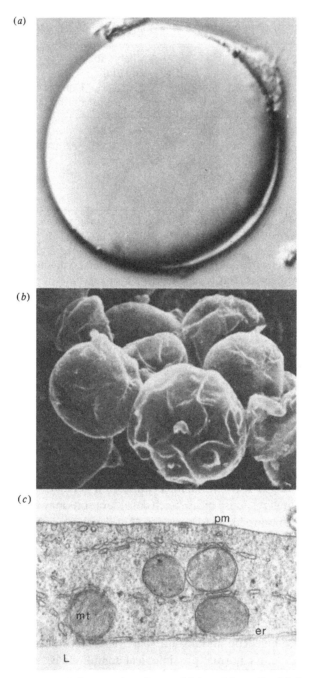

Fig. 5.11. Micrographs of rat epididymal fat cells. (*a*) Interference contrast micrograph obtained using the light microscope. The isolated cell, which is approx 50μm in diameter, is almost filled by

tion falls, triggering a rise in plasma glucagon concentration and fall in plasma insulin concentration that together cause an increase in lipolysis. This can result in a net loss from adipose tissue of triacyl glycerol per day of 150–200 g, sufficient to provide the calorie requirement of 1600 kcal per day. In normal subjects, adipose tissue stores of triacyl glycerol are sufficient to maintain adequate fatty acid supply to other tissues for more than 40 days (sometimes more than 60 days) of starvation.

Triacyl glycerol is stored mainly as a central lipid droplet surrounded by a thin cytoplasmic rim in the cells of white adipose tissue (Fig. 5.11). From many mammalian species, individual adipocytes are easily prepared by collagenase digestion of adipose tissue pieces. The resulting cell suspensions have proven an ideal experimental system for the study of lipolysis, since there is no barrier to the free access of hormones to the cell surface. The evidence that fast-acting lipolytic hormones act via a cyclic AMP-mediated mechanism was established using the criteria discussed above, i.e. the time course of increase in cyclic AMP concentration (Fig. 5.12), the effects of dibutyryl cyclic AMP and the effects of inhibitors of cyclic AMP phosphodiesterase. The rate-limiting enzyme in triacyl glycerol lipolysis is a hormone-sensitive lipase that can remove fatty acids stepwise to form sequentially diacyl glycerol, monoacyl glycerol and glycerol. The released free fatty acids bind to albumin in the circulation and are transported to other tissues. The lipase is not the easiest enzyme to study *in vitro*, since its substrate, triacyl glycerol, is immiscible with water. However, it has been possible to show that activation by cyclic AMP is mediated through cyclic AMP-dependent protein kinase (as in Fig. 5.9) causing phosphorylation of a specific serine of the lipase.

Experiments *in vitro* have shown that lipolysis in mammalian adipocytes is stimulated by a wide variety of catecholamine and peptide hormones. Consideration of *in vivo* concentrations of these hormones and the relative importance of circulating catecholamines or the effect of nervous stimulation have suggested that the normal *in vivo* stimulus of lipolysis is noradrenalin released by sympathetic nerve endings. In

Fig. 5.11. (*cont.*)

the central lipid droplet. The cytoplasm forms a thin rim except in the region of the nucleus, giving the cell its 'signet ring' shape. (*b*) Scanning electron micrograph showing the smooth surface of rat epididymal fat cells. (*c*) Transmission electron micrograph of fat cell cytoplasm showing the central lipid droplet (L), plasma membrane (pm), mitochondria (mt) and an extensive network of smooth endoplasmic reticulum (er).

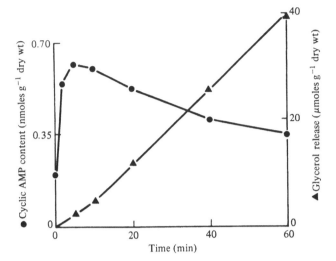

Fig. 5.12. Time course of the hormonal stimulation of cyclic AMP content and triacylglycerol breakdown in isolated rat fat cells. Equal aliquots of isolated cells were incubated with albumin in a buffered salt solution. Adrenalin $(5.5\,\mu M)$ was added at the start of the experiment and the incubation stopped at various times. Whereas cyclic AMP content reached a maximum within 5 min, a steady rate of glycerol release was established by 10 min and maintained for 60 min.

starvation, circulating glucagon is also likely to be important as a stimulating hormone. Lipolysis is therefore controlled by more than one stimulatory route. In fact, the concentration of the antilipolytic hormone insulin is also of importance in determining the rate of lipolysis since it can inhibit sub-maximal rates of hormone-stimulated lipolysis. The mechanism of this insulin effect is not clear because although insulin may reduce raised cyclic AMP concentrations by inhibiting adenylate cyclase and stimulating the phosphodiesterase, the level of reduction may not be sufficient to account for the lower rate of lipolysis (Fig. 5.13). Despite the difficulty in understanding the biochemical mechanism of the insulin effect, the control of lipolysis demonstrates the way in which the complex interplay of several hormones and of the endocrine and nervous systems can combine to control a metabolic pathway in different physiological states.

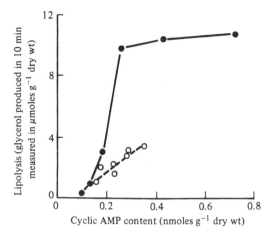

Fig. 5.13. The relationship of lipolysis to cyclic AMP concentration in rat fat cells. Cells were incubated with various concentrations of adrenalin (0.05–$5.0\,\mu$M) and the rate of lipolysis plotted against cyclic AMP concentration measured 5 min after addition of hormone (●). Maximum lipolysis was obtained with a 2.5-fold increase in cyclic AMP concentration. In the presence of the same adrenalin concentrations plus 0.35 nM insulin, cyclic AMP concentration could be increased fourfold without lipolysis reaching half maximum rates (○). Data from Dr K. Siddle.

The control of glycogen breakdown and synthesis

Glycogen is the major storage form of carbohydrate in the mammal. It is a polymer formed of α 1–4 linked glucosyl units also containing some α 1–6 linked units to form branches. It is found mainly in liver and muscle (approximately 50 % in each) and is a convenient storage form of glucose, since its high molecular weight ensures that storage entails no detrimental osmotic consequences for the cell. In muscle, glycogen serves as a source of phosphorylated glucose units to supply ATP for muscle contraction by anaerobic glycolysis, whereas in liver, it acts as a buffer to maintain the concentration of plasma glucose under fasting conditions to ensure the glucose supply to the brain. Glycogen can be converted to glucose only in the liver and not in muscle, since the latter lacks the enzyme glucose 6-phosphatase. In conditions of high plasma glucose concentration (i.e. after feeding), the plasma insulin concentration rises and causes glycogen synthesis to be stimulated. This occurs in both tissues as a result of activation of glycogen synthase. In addition, in muscle, there is stimulation of glucose uptake across the plasma membrane. The mobilization of glycogen is stimulated by glucagon which promotes glycogenolysis in the liver but not in skeletal muscle, the catecholamine noradrenalin released by sympathetic nerve endings which acts likewise (since sympathetic innervation is absent from skeletal muscle) and the catecholamine adrenalin which promotes glycogenolysis in both tissues. The protein phosphorylation and dephosphorylation mechanisms by which these three agents and insulin interact to affect the rate-limiting enzymes responsible for glycogen synthesis and degradation are shown in Fig. 5.14. Although the system is complex, the multi-step control sequence illustrated has advantages other than amplification of hormone signals. It allows the coordinated control of synthesis and degradation by the simultaneous activation or inhibition of enzymes with potentially opposing functions. It also enables non-hormonal control to be superimposed at various levels in the cascade sequence, allowing responses to be finely tuned to suit a variety of different physiological situations. This is particularly necessary since the same metabolic pathways are responsible for glycogen synthesis and degradation in muscle and liver yet the glycogen in each fulfils a different physiological role. Whereas in liver during fasting, glycogen normally lasts for about 24 h, muscle glycogen reserves can be exhausted after 1–2 min of exercise. The activation of glycogen synthesis in muscle after exercise is a slower process, glycogen synthase activity rising over a period of 30 min and then declining as glycogen stores are replenished.

The mechanisms shown in Fig. 5.14 have been established largely as a result of the careful purification and analysis of the various enzymes involved, especially from skeletal muscle. These studies have provided the following major concepts about the glycogen breakdown and synthesis pathways and their control.

1. *The activation of phosphorylase (the enzyme breaking down glycogen to glucose 1-phosphate), by conversion from the* b *to* a *form, is accompanied by its phosphorylation.* This occurs on a serine residue using ATP as substrate. Phosphorylase *b* is a dimer of two identical Mr 100 000 subunits, and phosphorylase *a* is a tetramer, although the polymerization has been shown to be of no importance to function since it occurs more slowly than activation of enzyme activity. Phosphorylase activity may also be controlled by the concentration of the metabolites glucose 6-phosphate (G6P), ATP, AMP and inorganic phosphate (P_i). AMP and P_i are synergistic activators of phosphorylase *b*, and the activation by AMP is antagonized by ATP and G6P. Phosphorylase *a* is almost fully active in the absence of AMP at

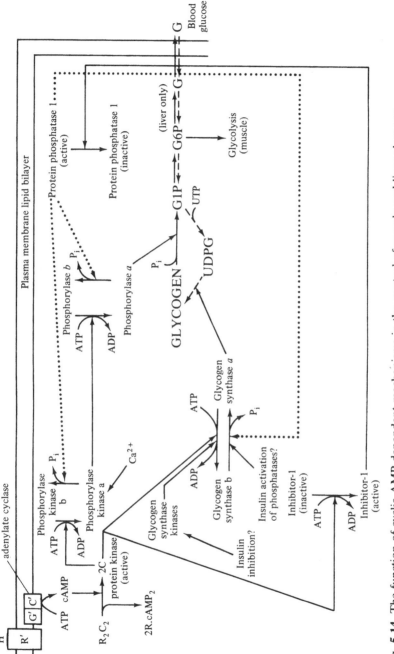

Fig. 5.14. The function of cyclic AMP-dependent mechanisms in the control of muscle and liver glycogen metabolism. H, hormone; R′, hormone receptor; G′, G$_s$ protein; C′, catalytic subunit of adenylate cyclase; R and C subunits of cyclic AMP-dependent protein kinase; G, glucose; G6P, glucose 6-phosphate; G1P, glucose 1-phosphate; P$_i$, inorganic phosphate; *a* is the more active and *b* the less active form of various enzymes.

saturating substrate concentrations and is not inhibited by ATP or G6P. While it is probable that the metabolite ratio AMP + P_i/ ATP + G6P is important in the control of glycogen breakdown (e.g. activating muscle glycogenolysis during contraction when AMP concentrations rise and ATP falls; inhibiting when there is an adequate supply of glucose and G6P), this has been difficult to prove since the *in vivo* intracellular concentrations of these metabolites have been difficult to determine. Independent evidence suggesting the importance of metabolite control comes from investigation of *I* strain laboratory mice which lack phosphorylase kinase yet are relatively healthy and can, for example, swim as long as normal mice. However, they cannot activate muscle glycogen breakdown as fast as normal mice, a deficiency which may be of importance in escaping from predators in the wild. Although changes in cellular energy state probably play a role in controlling glycogenolysis in muscle, they are unlikely to be important in liver. ATP and AMP levels in liver do not change significantly during the first 24 h of fasting.

2. *The enzyme phosphorylase kinase which activates phosphorylase can itself be activated as a result of serine phosphorylation by cyclic AMP-dependent protein kinase.* Phosphorylase kinase is a complex enzyme consisting of four subunits, α, β, γ and δ, the γ subunit being catalytically active. Changes in enzyme activity correlate with phosphorylation of the β subunit. While this may seem an adequate explanation of hormonal activation, it was not until the discovery of the δ subunit that a more complete understanding of activation of phosphorylase kinase associated with the onset of muscle contraction was obtained.

The δ subunit is identical to a protein called calmodulin that is closely related to troponin C, the protein conferring Ca^{2+} sensitivity to the contractile apparatus in muscle. Both calmodulin and troponin C can bind up to four Ca^{2+} ions per mole at μM concentrations of Ca^{2+}. Binding of Ca^{2+} to the δ subunit is a prerequisite for activation of the γ subunit of phosphorylase kinase. In fact, phosphorylase kinase interacts with a second molecule of calmodulin termed δ' which can activate dephosphorylated phosphorylase kinase *b*. Troponin C may substitute for δ' in this activation and may be important *in vivo*. The importance of the interaction of phosphorylase kinase with Ca^{2+} binding proteins is that it helps account for the activation of the enzyme when intracellular Ca^{2+} concentrations rise to the μM

range as happens in muscle during contraction. Overall, phosphorylase kinase *b* can be activated 20- to 30-fold by troponin C at μM Ca^{2+}, but a further 15-fold by phosphorylation and conversion to phosphorylase *a*. Both *b* and *a* forms of the enzyme have an absolute requirement for Ca^{2+}.

Activation of phosphorylase kinase by a rise in intracellular Ca^{2+} concentration may also be important in the activation of liver glycogen breakdown by adrenalin. Although, in the initial experiments with dog liver, this hormone was shown to act via binding to β receptors causing a rise in cyclic AMP concentration, it appears that in some species, notably the rat, its effects on liver glycogen breakdown are mediated by α_1 receptors causing an increase in cytosolic free Ca^{2+} concentration.

3. *The activated* a *forms of phosphorylase and phosphorylase kinase can be converted to the inactive* b *forms by the action of protein phosphatases.* The major phosphatase termed protein phosphatase-1 is itself controlled by an inhibitor protein that is an active inhibitor when phosphorylated. This phosphorylation is mediated by cyclic AMP-dependent protein kinase. Thus, when intracellular cyclic AMP concentrations are raised, cyclic AMP-dependent protein kinase, phosphorylase kinase, phosphorylase and inhibitor protein are active and protein phosphatase inactive. The converse is true when cyclic AMP concentrations are low.

4. *Glycogen synthesis is activated when glycogen breakdown is inactivated and vice versa.* The rate-limiting reaction in glycogen synthesis is the transfer of glucosyl residues from uridine diphosphate glucose (UDPG) to glycogen, catalysed by glycogen synthase. This enzyme exists in an inactive *b* form when phosphorylated and an active *a* form when dephosphorylated. Phosphorylation is complex and can occur on many serine residues at seven sites in the molecule, though one particular site, site 3, is thought to be most important in controlling activity. This site, which contains three serine residues, is phosphorylated by a specific kinase called glycogen synthase kinase-3 (GSK-3). Effects of phosphorylation at other sites may be additive to the effect at site 3. Thus, cyclic AMP-dependent protein kinase which phosphorylates sites 1 and 2 may play a role in inactivation of glycogen synthase. Removal of phosphate from the serine residues at site 3 activates the enzyme and this may be stimulated by insulin. This dephosphorylation may be mediated by protein

phosphatase-1 which in turn is under the control of the inhibitor protein described above. Since the inhibitor protein/protein phosphatase-1 system may be modulated by cyclic AMP-dependent protein kinase, this also provides a means of switching off glycogen synthesis when glycogenolysis is in progress.

Like phosphorylase, glycogen synthase is subject to control by the concentration of intracellular metabolites and this may add fine tuning to the system of hormonal control. Studies *in vitro* have shown that the phosphorylated *b* form of glycogen synthase may be fully activated by G6P, whereas the dephosphorylated *a* form is fully active in its absence. Activation of the *b* form by G6P is antagonized by ATP and P_i. The *a* form may also be inhibited by ATP, though this inhibition is reversed by very low concentrations of G6P. The significance of these effects *in vivo* is not obvious, since, for example at the concentrations of these metabolites thought to exist in resting muscle, the *b* form should be inactive and the *a* form fully active.

Despite the complexities of the control processes described above and the elegant ways in which they interact, it is likely that further study of glycogen degradation and synthesis will produce more surprises about the mechanisms of hormonal control. Other protein phosphatases and phosphatase inhibitors certainly exist within mammalian cells, differing from those described above. Their study should certainly help with the central remaining difficulty, the mechanism of activation of glycogen synthase by insulin (inhibition of GSK-3 or activation of a site 3 phosphatase).

Other second messengers

The discovery that some hormones mediate their effects by raising the intracellular concentration of cyclic AMP led to an extensive and often fruitless search for other second messengers in cases where alterations in intracellular cyclic AMP concentrations were not observed. Other cyclic nucleotides were obvious candidates and in the case of one, cyclic GMP (with guanosine replacing adenosine in the nucleoside), a guanylate cyclase, cyclic GMP phosphodiesterase and cyclic GMP-dependent protein kinase were shown to exist in mammalian cells. Although a wide variety of hormones have been shown to alter the cyclic GMP content of various cells, it has not been possible to satisfy fully the criteria for cyclic GMP being a second messenger in any single case. There is, however, a growing body of evidence that cyclic GMP plays a role in a wide variety of physiological events including cell proliferation and vision.

An alteration of intracellular Ca^{2+} ion concentrations has long been regarded as a possible mediator of hormone action. Not only has an increase in cytosolic Ca^{2+} concentration been shown to mediate muscle contraction and, more recently, activation of phosphorylase kinase, but it is also important in some cells as a factor coupling secretion to defined stimuli (see Chapter 6). The resting cytosolic Ca^{2+} concentration in mammalian cells is considered to be of the order of $0.1\,\mu M$, considerably less than the extracellular calcium concentration of approximately 1 mM. Total cell Ca^{2+} concentration is also about 1 mM, the vast bulk being sequestered in intracellular organelles (endoplasmic reticulum and mitochondria), and bound to membranes or cell proteins. Thus, only a small amount of extracellular Ca^{2+} or intracellular sequestered Ca^{2+} needs to be released to have a considerable impact on cytosolic Ca^{2+} concentration. Moreover, the existence of calmodulin as a ubiquitous protein that binds Ca^{2+} in the μM range and can activate several enzymes when Ca^{2+} is bound provides a mechanism allowing the amplification of changes in cytosolic Ca^{2+} concentration to modulate metabolic pathways. The major problem in defining Ca^{2+} as a second messenger has been the difficulty in accurately determining cytosolic concentrations. In recent years, fluorescent Ca^{2+}-sensitive indicators have been developed that can cross the plasma membrane as lipid-soluble esters which are then hydrolysed allowing the indicator to 'report' on alterations in cytosolic Ca^{2+} concentration. In this way, several hormones have been shown to affect cytosolic Ca^{2+} concentration, a first essential step in determining whether it can act as an intracellular messenger.

Recently, it has been shown that a number of hormones and growth factors can cause an increase in cytosolic Ca^{2+} concentration as a result of the effects of inositol triphosphate (IP_3), a product of the breakdown of inositol-containing phospholipids. These occur in most membranes and can contribute up to 10 % of the total lipid. It has been known since the 1950s that a variety of hormones, neurotransmitters and alterations in growth conditions can stimulate the degradation and resynthesis of phosphatidyl inositol. More recently, the derivative phosphatidyl inositol 4,5-biphosphate (PIP_2) has been identified as the metabolically active fraction (10 % of total inositol phospholipids) which is broken down to IP_3 and DAG (Fig. 5.15). Using permeabilized cells, it has been shown that inositol triphosphate added extracellularly can enter the cells and cause release of Ca^{2+} from intracellular stores. Thus it may act as a second messenger with Ca^{2+} strictly being a third messenger. The second product of PIP_2 inositol degradation, diacylglycerol (DAG), may also be a second messenger since it activates a Ca^{2+}-dependent enzyme, protein kinase C, by increasing its sensitivity to Ca^{2+} up to 1000-fold. Thus, Ca^{2+}

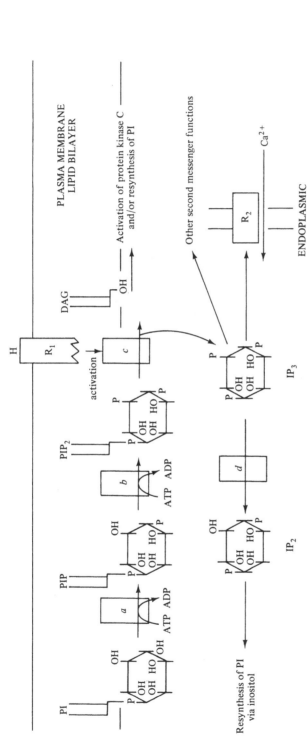

Fig. 5.15. The formation of inositol triphosphate and diacylglycerol. Hormones (H) or other agonists bind to membrane receptors (R_1) activating a phosphodiesterase (c) which hydrolyses phosphatidyl inositol 4,5-biphosphate (PIP_2) to diacylglycerol (DAG) and inositol 1,4,5-triphosphate (IP_3). Both of these may be used in the resynthesis of phosphatidyl inositol (PI). IP_3 can bind to a specific receptor (R_2) on the endoplasmic reticulum causing the release of Ca^{2+} into the cytosol. IP_3 is inactivated by inositol triphosphatase (d) which converts it to inositol 1,4-biphosphate (IP_2). Other lipids and enzymes shown are phosphatidyl inositol 4-phosphate (PIP), PI kinase (a), and PIP kinase (b). The fatty acid side-chains of PI are predominantly stearic acid and arachidonic acid, respectively.

Table 5.3 *The major effects of insulin on its principal target organs.*

Effects	Principal target organs		
	Muscle	Liver	Adipose tissue
Short-term effects			
1 Stimulation of plasma membrane transport			
– glucose uptake	√		√
– amino acid uptake	√	√	√
– K^+ uptake	√	√	√
2 Stimulation of glycogen synthesis	√	√	
3 Stimulation of fatty acid synthesis		√	√
4 Inhibition of glycogen breakdown	√	√	
5 Inhibition of lipolysis			√
6 Inhibition of gluconeogenesis		√	
Longer-term effects			
1 Stimulation of protein synthesis	√	√	√
2 Stimulation of RNA synthesis	√	√	
3 Inhibition of protein degradation	√	√	√
4 Induction of synthesis of specific enzymes (e.g. tyrosine amino-transferase and glucokinase)		√	
Maintaining effects (revealed by long-term insulin deprivation)			
1 Maintenance of glucose transport	√		√
2 Maintenance of glucose phosphorylation	√		√
3 Glycogen synthesis		√	
4 Lipid synthesis		√	√

may potentially regulate metabolic pathways in at least two ways, via calmodulin or protein kinase C. It is important to realize that inositol triphosphate and diacylglycerol may act synergistically as second messengers, though the physiological effects and substrates of these agents remain to be fully established. It has been suggested that hormone receptor binding activates the phosphodiesterase responsible for hydrolysis of PIP_2 via a specific G protein in the plasma membrane, an interesting analogy with the adenylate cyclase system.

The mechanism of insulin action
Insulin is a polypeptide hormone with a multiplicity of both short-term and long-term effects (Table 5.3). It was isolated in the 1920s and since

then has been used successfully in the treatment of diabetics. Insulin was the first protein of which the primary sequence was established (in the early 1950s) and one of the earliest of which the complete three-dimensional structure was known by X-ray crystallography. Despite these achievements and the more recent cloning and sequencing of its plasma membrane receptor, the mechanism of action of insulin has remained extraordinarily elusive.

It was recognized early that a major effect of insulin was to stimulate the rapid uptake of blood glucose into peripheral tissues. For some time, it was thought that all insulin effects might be explained by the increased glucose transport, but the development of *in vitro* systems where insulin effects were observed in the absence of glucose confounded this hope. Even the stimulation of glucose transport by insulin has proven to be remarkably complex. The effect of insulin is due to an increase in the maximum activity (V_{max}) of plasma membrane glucose transport with little change in the affinity for glucose (K_m). Rather than a direct covalent change in the glucose transporter protein this V_{max} increase is apparently brought about by insulin stimulating the recruitment into the plasma membrane of extra transporter molecules from intracellular membrane sites. This has been shown by subcellular fractionation studies of both adipose tissue and muscle. In these experiments, the glucose transporter content of different membrane fractions was assessed by the extraction of glucose transporter molecules and their reconstitution into artificial phospholipid vesicles (liposomes) or by the binding of radioactively labelled drugs that specifically react with the glucose transporter.

Attempts to identify a classical intracellular second messenger for insulin have been largely unsuccessful. The obvious antagonism between insulin and hormones that stimulate adenylate cyclase in controlling processes such as glycogen metabolism and lipolysis suggested that insulin might act by decreasing tissue cyclic AMP content. In fact, insulin has been shown to be capable of lowering cyclic AMP concentration as a result of activation of a high-affinity cyclic AMP phosphodiesterase and inhibition of adenylate cyclase, the latter possibly occurring via an effect on a specific inhibitory G protein. Unfortunately, while lowering cyclic AMP concentration may play a role in some insulin actions under physiological conditions, it is clear that many effects of insulin (e.g. stimulation of muscle glucose uptake and glycogen synthesis) can occur without effects on cyclic AMP content. Moreover, as described above, even in adipose tissue, the observed lowering of a raised cyclic AMP content by insulin is insufficient to account for the magnitude of its antilipolytic effect in the presence of adrenalin (Fig. 5.13).

At different times, alternative second messengers including cyclic GMP, Ca^{2+} and products of phospholipid metabolism have been proposed as the insulin second messenger. Although changes in their intracellular concentration in response to insulin have been observed, other criteria for a second messenger have not been satisfied. It has also been proposed that the insulin second messenger may be a small peptide released from the plasma membrane after exposure to insulin and perhaps even consisting of a fragment of the hormone or its receptor. It is not clear that any such peptide can act as an insulin second messenger under physiological conditions, nor how it could achieve amplification of the hormonal signal.

In recent years, increasing emphasis has been placed on understanding the complex intracellular phosphorylations and dephosphorylations occurring on target enzymes as a result of insulin action, e.g. in controlling glycogen synthesis and degradation, as discussed above. Insulin effects on the phosphorylation state of intracellular proteins are complex. They often occur at only one or a few of the potential phosphorylation sites and they may be phosphorylations (e.g. ATP citrate lyase; acetyl CoA carboxylase) or dephosphorylations (e.g. mitochondrial pyruvate dehydrogenase α subunit, glycogen synthase). The identity of enzyme(s) mediating insulin-sensitive changes in phosphorylation is unknown but may include kinases, phosphatases and inhibitor proteins. It has been suggested that insulin binding to its receptor may activate a membrane protein kinase which, perhaps following dissociation from the membrane, can phosphorylate many proteins providing the necessary amplification of the hormonal response.

The most exciting recent advances in attempts to understand the mechanism of insulin action have come from the study of the insulin receptor (Fig. 5.16). This is a disulphide bonded tetramer consisting of two dimers each containing two distinct polypeptide chains, α (Mr 130 000) and β (Mr 95 000), which are derived from a single precursor. The extracellular α chain is glycosylated and can be chemically cross-linked to insulin, showing that it forms part of the insulin binding site, although affinity labelling studies suggest the β chain is also involved. The transmembrane β chain is autophosphorylated on tyrosine residues as a result of insulin binding both *in vitro* and in intact cells. This autophosphorylation of the receptor provides an obvious link with the ability of insulin to modulate the phosphorylation state of intracellular target enzymes. However, it has been possible only to show that the receptor is a tyrosine kinase, yet it appears that modulation of intracellular enzyme activity occurs via serine phosphorylation. The possibility of a receptor-

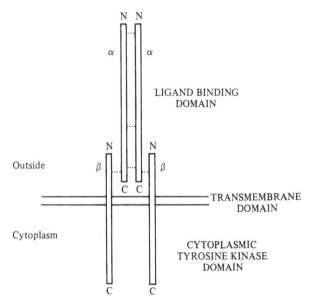

Fig. 5.16. The insulin receptor consisting of two extracellular α chains and 2 transmembrane β chains. The disulphide bonded $\alpha\beta$-dimers are themselves joined by disulphide bonds (possible regions of the molecule involved in disulphide bonding are shown ---). There are approximately 10^5 such receptors on a liver cell and the maximum effect is obtained when only 10 % have bound insulin.

associated serine-specific kinase being lost or inactivated during receptor purification has been suggested.

The study of isolated as well as cell surface insulin receptors is providing a number of additional clues about insulin action. Anti-insulin receptor antibodies are able to mimic the effects of insulin including receptor phosphorylation. Only bivalent antibodies, not their univalent fragments, are active. This may indicate *in vivo* that receptor clustering and endocytosis occur coincidentally with receptor activation. While it is not obvious that receptor internalization is necessary for acute insulin action, it is intriguing that insulin-stimulated glucose transport apparently requires movement of membrane proteins. Moreover, it is known that prolonged challenge of cells with insulin leads to net loss of receptors from the plasma membrane (down regulation) and it may also provide a route of signalling to the cell interior for the long-term effects of insulin. Finally, a knowledge of primary sequence of the insulin receptor derived from molecular cloning studies has shown a sequence similarity between the cytoplasmic portion of the β chain and the tyrosine-specific protein kinase domains of a number of oncogene products, suggesting an important role for this kinase activity in cell proliferation responses.

Long-term effects of hormones (steroid hormone action)

Although some acute-acting hormones which bind to a cell surface receptor also have long-term actions, it is not obvious how these latter effects are mediated. However, in the case of steroid hormones which have a variety of long-term effects caused by activating specific gene expression, a clearer view of the mechanism of action is possible. Steroid hormones exert their major effects on target tissues by a two-step process (Fig. 5.17). In the first step, the hormone is selectively accumulated in target cells because these cells contain hormone-specific, high-affinity ($K_D \, 10^{-9}$–10^{-10}M) cytoplasmic receptor proteins. In the second step, the hormone-receptor complex moves into the nucleus, interacts with the chromatin and causes changes in specific mRNA synthesis. The specific events of the two-step hypothesis have been demonstrated, for example, by examining the effect of oestradiol (an oestrogen, Table 5.1) on uterine cells, but some difficulties with the hypothesis remain, as follows.

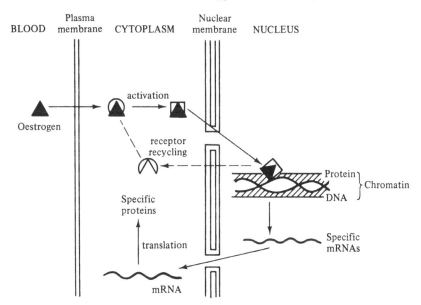

Fig. 5.17. The two-step model for the action of an oestrogen on its target cell. The oestrogen binds to a cytoplasmic receptor which is activated to pass into the nucleus and interact with chromatin. This results in the production of specific messenger RNAs which enter the cytoplasm and are translated to provide the overall tissue response to the oestrogen.

1. *The mechanism of transfer of steroid hormone from plasma to target cell occurs by passive diffusion.* In normal women, only about 2 % of plasma oestrogen is free hormone, the remainder being bound to albumin and sex steroid-binding globulin. The free hormone is regarded as the biologically active component and most evidence indicates that it enters cells by passive diffusion, accumulating in target cells as a result of binding to cytosolic receptor proteins. Some research workers have suggested facilitated transport of oestrogen across the plasma membrane, and others have suggested that plasma protein-bound steroid is internalized by cells before hormone release.

2. *Cytoplasmic steroid receptors are activated by steroid binding.* Steroid receptors have been identified by binding of radioactive steroid with subsequent characterization by density gradient centrifugation. Thus, rat uterine cells have approximately 20 000 oestrogen receptors per cell with a sedimentation coefficient of 4S measured in high ionic strength buffer. Each oestrogen receptor is thought to be a single polypeptide chain of Mr 80 000 with a single hormone binding site (other steroid hormone receptors have been found to be heterodimers, each monomer capable of binding hormone). It was thought that hormone binding allowed translocation to the nucleus by a temperature-dependent (prevented at 0–4 °C) activation process reflected both by an increase in sedimentation coefficient (4S to 5S) and by the ability to bind to isolated nuclei. However, it has been suggested recently that unfilled receptors normally exist in the nucleus and are found only in cytosolic fractions as a result of nuclear damage during homogenization protocols. Attempts to localize steroid hormone receptor proteins using monoclonal antibodies and immunomicroscopic techniques support this conclusion. Although activation of receptor is thought to occur as a result of steroid binding, the mechanism is still not well defined. Binding of steroid hormone-receptor complex to many nuclear components has been reported, although most interest is now focussed on interactions with the DNA of the activated genes. Gene transfer experiments in which defined genes and upstream sequences are introduced into cells containing steroid receptors but not normally expressing the gene have suggested the importance of multiple DNA binding sites both upstream of and within the target gene. It is suggested that these sites act as enhancer sequences for transcription when the steroid hormone-receptor

complex is bound. It remains possible that nuclear non-histone proteins also play a role in steroid hormone activation of gene expression.

3. *Switching off steroid hormone action may require proteolytic cleavage of the steroid hormone receptor.* It is not clear how reversal of steroid hormone activation of cells occurs. There is no evidence for specific breakdown of steroid hormones in target cells. However, exposure of target cells to high concentrations of steroid hormone can cause greater than 50 % depletion of cytoplasmic receptor activity. This receptor does not appear in the nucleus and recovery of cytoplasmic receptor activity can be blocked by inhibitors of protein synthesis.

Although the mechanisms by which steroid hormones exert their effects are still poorly understood compared to the actions of many acute-acting polypeptide hormones, a more complete picture should emerge as the overall mechanisms of control of eukaryotic gene expression become clearer.

Hormone–hormone interactions (thyroid disease)

In many cases, physiological responses to hormone action are responses to the presence of more than one hormone. Examples of the interaction of acute-acting hormones to control lipolysis and glycogen metabolism were given above, e.g. insulin antagonizing the actions of glucagon and adrenalin. Examples of long-acting hormones affecting tissue responses to acute-acting hormones are also known, e.g. the sensitivity of adipocytes to lipolytic hormones depends on thyroid status and glucocorticoids are permissive to the effects of glucagon on liver glycogen metabolism. Long-acting hormones may also interact to produce a given response. Thus some effects of the steroid hormone progesterone only occur in tissues previously exposed to oestrogens, presumably due to the oestrogens promoting the formation of progesterone receptors.

A further complexity of endocrine systems is that the release of one hormone may be controlled by the action of another (see Chapter 6 for endocrine control of insulin secretion). A particularly clear example and one with considerable medical significance involves the control of thyroid hormone secretion. The thyroid gland produces two biologically active hormones, 3,5,3',5'-tetraiodothyronine (thyroxine or T4) and 3,5,3'-triiodothyronine (T3, Table 5.1). These are present in the bloodstream at concentrations of about 10^{-7}M for T4 and 10^{-9}M for T3, both hormones being mostly bound to a series of serum-binding proteins. The thyroid hormones affect the growth, development and metabolic activity of all cells. T3 is three to five times more active than T4, which probably acts as a precursor of T3. The thyroid hormones appear to have specific nuclear (though not cytosolic) receptors and may act to regulate gene transcription in a manner analogous to steroid hormones.

The production of T3 and T4 by the thyroid gland is under the control of thyrotropin (TSH, Table 5.1), which binds to a thyroid cell surface receptor activating adenylate cyclase. Intracellularly, a series of enzymatic reactions converting trapped iodide to thyroid hormones are regulated. TSH is secreted by cells in the anterior pituitary and if the serum concentrations of T3 and T4 fall, TSH secretion is increased; conversely, if serum T3 and T4 rise, then TSH secretion is inhibited. TSH synthesis and secretion by pituitary cells is stimulated by a tripeptide, TRF (Table 5.1) produced by peptidergic neurones in the hypothalamus. T3 and T4 probably block the action of TRF on the pituitary to prevent TSH secretion (Fig. 5.18).

Diseases due to abnormal thyroid function are relatively common among the general population, affecting perhaps as many as 1–2 % of people in the UK (Table 5.4). Both hypo- and hyperthyroidism are 10 times more common in women than in men and, if not treated, can

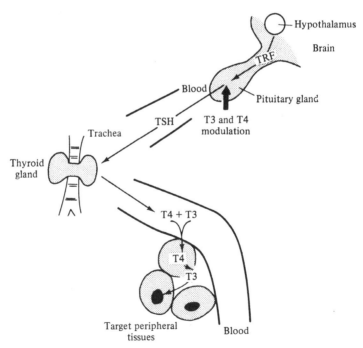

Fig. 5.18. Schematic diagram of the interaction of hormones involved in controlling thyroid hormone secretion.

Table 5.4 *Symptoms of thyroid disease*

Disease	Symptoms
Hypothyroidism (myxoedema)	Lack of energy; inability to tolerate cold weather; slow pulse rate; weight gain; dry, thickened skin, cool to touch; hair dry and brittle; characteristically outer third of each eyebrow lost; prolonged relaxation phase of knee-jerk response; slow to respond to questions; sometimes thyroid gland enlarged (goitre). In advanced cases, hallucinations can appear (myxoedema madness). If the deficiency is present from birth (neonatal hypothyroidism), mental retardation and dwarfism result (cretinism).
Hyperthyroidism (Graves' disease)	Intolerant of hot weather; anxiety; palpitations; rapid pulse rate; weight loss; muscle weakness (thyrotoxic myopathy); warm moist skin; tendon jerks hyperactive; pop-eyed appearance due to eyeballs protruding and eyelids retracting.

ultimately be fatal. Abnormal serum levels of T3 and T4 can arise either from a defect in the thyroid gland itself or from abnormal secretion of TSH by the pituitary. In fact, pituitary causes of thyroid disorders are uncommon and in the great majority of cases the thyroid gland is at fault. The most useful biochemical tests in thyroid disease are the measurements of T3 and T4 in the circulation (both of these are decreased in hypothyroidism and raised in hyperthyroidism, whether caused by a pituitary or a thyroid defect) and the measurement of TSH. If the hypothyroidism is of thyroid origin, the TSH level is raised (in effect to try to stimulate the defective gland); if, however, the hypothyroidism is due to a pituitary defect (which is rare), then the TSH level is undetectable. If hyperthyroidism is due to thyroid overfunctioning, then the TSH level is very low because of the feedback of the raised T3 and T4 on the pituitary. Lastly (and rarely), if hyperthyroidism is due to a pituitary over-production, TSH is high along with serum T3 and T4. In practice, the measurement of TSH is most useful in diagnosing hypothyroidism due to primary thyroid failure (the TSH level being raised) and in screening neonates for cretinism (estimated incidence 1 in 3500 live births) when very high TSH levels are normally found. Approximately 10 000 measurements of TSH are done in one of our own hospitals (Southampton General) alone each year, demonstrating both the value of immunoassay techniques (Chapter 3) and of a knowledge of hormone interactions.

6

Secretion

The variety of secretory processes

All cells release metabolic products into the surrounding milieu and these may be taken up by other cells or bind to cell surface receptors enabling modulation of target cell function. At the simplest level, the release of metabolic products may involve only transport across the plasma membrane, though in many eukaryotic cells a variety of compounds are pre-packaged in membrane bound vesicles and released as a result of fusion of these vesicles with the plasma membrane. This release process of pre-packaged materials is known as emiocytosis or exocytosis. The range of materials released from mammalian cells by exocytosis is considerable and includes neurotransmitters from nerve terminals, a large number of hormones from endocrine cells, digestive enzymes from exocrine cells, blood plasma constituents such as albumin from liver cells and immunoglobulins from plasma cells, and histamine, which mediates local inflammatory reactions, from mast cells. Compounds released may be polypeptides synthesized on the endoplasmic reticulum and packaged by the Golgi apparatus into secretory vesicles or small molecules synthesized in the cytoplasm and transported into pre-existing vesicles. In both cases, final chemical modification of the secreted material may occur in the secretory vesicle or even after exocytosis and release into the surrounding medium.

It is clear that, despite the variety of materials exocytosed by cells, there are common underlying features in the mechanisms by which secretory vesicles are formed and in the biochemical events involved in the physiological control of secretion. Several of these features are described below, followed by a more specific discussion of the mechanism of insulin synthesis and secretion. In addition to having many characteristics in common with other secretory systems, deficiencies of the insulin secretory pathway lead to diabetes, a major disease, the possible origins of which are also discussed.

Constitutive and triggered pathways of secretion

Mammalian cells use two kinds of pathway for exocytosis. The first, a constitutive pathway, is not acutely regulated by physiological stimuli and accounts for such processes as mucus secretion by intestinal goblet cells and albumin secretion by hepatocytes. The second pathway is a triggered pathway in which secretory vesicles filled with the material to be secreted accumulate in the cytoplasm and are triggered to fuse with the plasma membrane as the result of some extracellular stimulus. Constitutive and triggered pathways may co-exist in the same cell type. A particularly clear example has been demonstrated in a cultured pituitary tumour cell line that synthesizes adrenocorticotrophic hormone and stores it in secretory granules (Fig. 6.1). The same cultured cells also produce an endogenous leukaemia virus and it appears that processing of viral coat proteins and of adrenocorticotrophic hormone occurs in similar or identical Golgi compartments. Isolated adrenocorticotrophic hormone secretory granules contain little viral coat glycoprotein and this fact plus considera-tion of pulse-chase labelling experiments has shown that, after leaving the Golgi apparatus, the viral envelope glycoprotein is transported to the cell surface by different vesicles on a faster route that is not sensitive to agents stimulating secretions of the hormone. Secreted by this constitutive route is not only viral coat glycoprotein but also some of the precursor protein

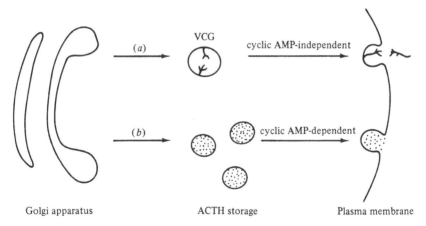

Golgi apparatus   ACTH storage   Plasma membrane

Fig. 6.1. A schematic model of the constitutive (*a*) and triggered (*b*) secretory pathways in cultured pituitary tumour cells. Adrenocorticotrophic hormone (ACTH) is stored in secretory vesicles and release is greatly stimulated by agents raising intracellular cyclic AMP concentration. Viral coat glycoprotein (VCG) moves by a rapid constitutive pathway to the cell surface. Modified from Gumbiner, B. & Kelly, R. B. (1982). *Cell*, **28**, 51–9.

for adrenocorticotrophic hormone, presumably leaking from the Golgi apparatus into this pathway rather than being processed and packaged in secretory granules. It should be noted that constitutive pathways of secretion may be the same pathways of intracellular membrane movement responsible for insertion of membrane components into the plasma membrane.

With a constitutive pathway of secretion, there is essentially no concentration process within the secretory vesicles and no storage within the cell of material to be secreted. Secretion by constitutive pathways is continuous with a rapid rate ($t_\frac{1}{2} \simeq 10$ min) of translocation from the Golgi apparatus to the cell surface. In the case of triggered exocytosis, material is often stored in secretory granules for days before the cell is stimulated. Either a small proportion or almost all of these granules will release their contents depending on the cell type, the material to be secreted and the nature of the stimulus. An example where almost total release occurs is histamine secretion from mast cells. In cells from immunized animals, the receptors on the mast cell surface bind IgE antibodies against the specific antigen with which the animal was immunized. Addition of the antigen causes cross-linking of the receptors and, within 10 s, the cell starts releasing granule contents, degranulation being essentially complete within 1 min. Rather less release is observed in other secretory systems. For example, there are estimated to be 13 000 granules containing the intracellular reservoir of insulin in a pancreatic beta cell, and approximately 10 % will be secreted in each hour during active periods of secretion. Similarly, it has been estimated that, in response to a 2-min pharmacological depolarization of the synaptic membrane of the mammalian cholinergic nerve terminal, less than 1 % of total acetyl choline stores are released.

The possibility of releasing only a proportion of the total intracellular store of material in secretory granules raises the question of whether this is a proportion of the material in each granule or all the material from some granules. In general, it appears that the latter occurs, though in some cells this may result from compound exocytosis where granules fuse together before or as they fuse with the plasma membrane rather than single granule fusion with the plasma membrane. It is interesting to note that, in the case of synaptic transmission, a mechanism for the quantal release of transmitter was proposed from the electrophysiological observation of stepwise fluctuations in the resting potential of the post-synaptic membrane before the identification and isolation of synaptic vesicles containing chemical transmitters.

In addition to secretion of granule contents, an inevitable consequence of both constitutive and triggered exocytosis is the insertion of secretory granule membrane into the plasma membrane. In a variety of cells, there is now good electron microscopic cytochemical evidence that the granule membrane can be recovered specifically by an endocytic process. Thus, when the secreted material is a relatively simple, small molecule, the recovered granule may be re-filled and re-used for further exocytotic events.

The intracellular pathways of synthesis and secretion of proteins
Over a 20-year period from 1950, the work of Palade using the techniques of subcellular fractionation, pulse-chase radioactive labelling and electron microscopic autoradiography largely established the intracellular routing of proteins to be secreted. Palade and co-workers maximized incorporation of ^{3}H-leucine into secretory proteins in the guinea-pig pancreatic exocrine cell by treating the animals with a fasting/feeding cycle before their experiments. They then showed that, after a single intravenous injection of ^{3}H-leucine, at 5 min the label was localized mostly in the endoplasmic reticulum, at 20 min it appeared in the Golgi complex and by 1 h in zymogen granules. Chasing the radioactive label with non-radioactive leucine helped to confirm the time course of the appearance of label in the various subcellular compartments and showed the successive transfer of label between these compartments. From these and other experiments came the following findings:

1. Proteins for export are synthesized by polysomes that are bound to the rough endoplasmic reticulum.
2. Proteins destined for export are never found in completed forms in the cytoplasm. They are segregated immediately into the lumen of the rough endoplasmic reticulum.
3. Proteins are transported from the endoplasmic reticulum through the Golgi apparatus to secretory vesicles, where they remain stored until secretion.

In recent years, genetic verification of the secretion pathway in eukaryotic cells has been obtained using yeast as a model system. Yeast mutants defective in protein export accumulate secretory organelles, become more dense than wild-type cells and are therefore easy to isolate. A variety of mutant strains that accumulate exported proteins have been studied and categorized as mutants accumulating either a structure related to the endoplasmic reticulum, a structure related to Golgi, or secretory vesicles. Constructing double mutants and then determining the intracellular structure accumulating has confirmed the temporal order of the secretory pathway and will allow in the future the sophisticated dissection of the steps involved in this pathway.

The signal hypothesis

The problem of deciding how a protein is destined for export is now thought to be explained by the signal hypothesis which was discussed in Chapter 4 (see Fig. 4.29) with respect to explaining the synthesis of membrane proteins. The first evidence providing information on the nature and location of export information came from experiments on the synthesis of IgG light chains (see Chapter 2) in an *in vitro* cell-free translation system. It was found that the IgG light chain was synthesized initially as a larger precursor form with a peptide extension at the amino terminus. While this observation was significant, it was not until the development of an assay for protein transport across membranes during synthesis *in vitro* that the function of the amino-terminal peptide extension was shown. In a cell-free translation system to which microsomal vesicles had been added, it was possible to demonstrate that newly synthesized secretory proteins were found inside the vesicles using the criterion of protease resistance, i.e. they were resistant to protease attack unless the membrane barrier was destroyed. The protein in the microsomal vesicles was also found to be mature, i.e. it had no amino-terminal extension. If microsomal vesicles were added to the cell-free translation system after completion of protein synthesis, then neither transport into the vesicles nor processing of the protein was detected. The signal hypothesis was thus formulated whereby a protein destined to be secreted from cells is synthesized initially as a larger precursor with a 15 to 30 amino acid, hydrophobic extension at the aminoterminus. This peptide extension, called the signal sequence, was proposed to initiate binding of protein being synthesized to the endoplasmic reticulum and allow vectorial transport into its lumen. The signal sequence is removed by a specific protease, the signal peptidase in the endoplasmic reticulum, before synthesis is complete.

Although the signal hypothesis has been subject to amendment since being formulated, it has gained enthusiastic and widespread support and there is a consensus that it explains the synthesis of all secreted proteins and the mechanism by which they enter the lumen of the endoplasmic reticulum. The overall scheme of the signal hypothesis was shown in Fig. 4.25 as applied to the synthesis of membrane proteins, which require internal stop–transfer sequences to prevent transport of the complete protein molecule into the lumen of the endoplasmic reticulum. It has been shown in some *in vitro* systems that synthesis of the protein temporarily stops after production of the signal peptide when this binds to a cytosolic signal recognition particle (SRP) made up of protein and RNA constituents. The SRP locates the ribosome on the endoplasmic reticulum by binding to a specific docking protein. The signal peptide then interacts with a membrane pore which includes membrane proteins (ribophorins) and SRP is released to the cytosol. This allows translation to resume and the newly synthesized protein to enter the lumen of the endoplasmic reticulum. Thus the SRP effectively couples protein synthesis with the secretion pathway.

There are some exceptions to the simplest version of the signal hypothesis described above. Ovalbumin, the egg white protein, does not have an amino-terminal signal sequence yet is secreted in typical fashion. In fact, it appears to have an internal signal sequence that is analogous to the amino-terminal extension in other secreted proteins. Thus, experimentally it was found that ovalbumin can compete with other newly synthesized secretory proteins for transport into microsomal vesicles, although no signal sequence is subsequently cleaved from the molecule. Internal signal sequences are also required for newly synthesized proteins destined for mitochondria and for membrane proteins in which the polypeptide chain crosses the phospholipid bilayer more than once (see Chapter 4).

While the signal hypothesis explains how secretory proteins pass across the endoplasmic reticulum membrane from the cytosol, they still have to be packaged into secretory granules and may require additional post-translational modification. They remain enclosed in membrane when moving from the endoplasmic reticulum to the Golgi apparatus and passing through this apparatus. The exact mechanisms of movement are not known, but it is suggested that coated vesicles are involved, as in the case of the equivalent transfer of newly synthesized membrane proteins discussed in Chapter 4. It is also likely that protein coats may be necessary for the formation of secretory vesicles from the Golgi apparatus and an example, insulin granule maturation, is described below. Although protein-coated membrane vesicles clearly play a role on secretory pathways, there is dispute as to the nature of the coat. While the coat may be clathrin (see Chapter 4), studies in yeast have shown that deletion of the clathrin heavy chain gene had little influence on the rate of intracellular transport and secretion of newly synthesized invertase. At least for this constitutive pathway in yeast, clathrin does not appear to be necessary for exocytosis.

There is some evidence that mutation in the protein to be secreted can affect transfer of the newly synthesized protein from the endoplasmic reticulum to the Golgi apparatus. The lung disease emphysema can be caused by a deficit of the serum proteinase inhibitor α_1-antitrypsin (see Chapter 1). This protein is synthesized in the liver and, in the case of the *Z* variant, a single glutamate–lysine point mutation results in aggregation of *Z* α_1-antitrypsin in the hepatocyte endoplasmic reticulum. Thus, only about 15 % escapes to be packaged for secretion. The remaining aggregated *Z*-antitrypsin in the endoplasmic reticulum is removed by intracellular proteolysis. The aggregation of *Z*-antitrypsin is accompanied by a distortion of hepatocyte architecture that helps explain why the *Z* variant, unlike other genetic antitrypsin deficiencies, is associated with liver disease as well as lung disease.

During maturation through the Golgi apparatus, secretory proteins are often modified by glycosylation reactions and finally must be concentrated during formation of secretory vesicles. Concentration is presumably more important where secretory vesicles are to be stored on triggered secretory pathways and occurs in terminal stages of passage through the Golgi apparatus. Electron microscopic evidence indicates that the exact location of the concentration step within the structure of the Golgi apparatus varies in different cell types. Full details of the biochemical mechanisms of concentration remain to be established.

The intracellular sorting of proteins to be secreted

The initial sorting of secretory from other proteins is achieved during synthesis by compartmentalization into the lumen of the endoplasmic reticulum as described above. However, it is clear that, even on constitutive pathways of secretion, newly synthesized proteins may be secreted at different rates. Using pulse-chase labelling experiments, studies on the secretion of a variety of proteins by mammalian liver cells have suggested that the major selection process occurs at the level of transfer from the endoplasmic reticulum to the Golgi apparatus. This transfer is energy-dependent and occurs via coated vesicles budding off the endoplasmic reticulum and fusing with the Golgi apparatus. Whether specific receptors are required to sort proteins from the endoplasmic reticulum into secretory pathways is not known nor are the features of primary or secondary structure they might recognize. However, it has recently been observed that a specific sequence of four amino acids (lys–asp–glu–leu; KDEL) at the C-terminal end of proteins in the lumen of the endoplasmic reticulum is responsible for these proteins being concentrated at this intracellular location. When the C-terminal amino acids are deleted using recombinant DNA techniques, the proteins are secreted; when added to normally secreted proteins, these are retained in the endoplasmic reticulum.

Once in the Golgi apparatus, passage through it on constitutive secretory pathways occurs at the same or very similar rates for different proteins. Experimentally, in a cultured pituitary tumour cell line, it has been shown that adrenocorticotrophic hormone can be diverted from a triggered to a constitutive secretory pathway using the weak base chloroquine which can destroy transmembrane pH gradients. This has suggested that acidity differences between organelles such as the Golgi apparatus and secretory granules may play a role in sorting proteins into such granules. Perhaps the protein to be secreted binds to a membrane receptor in the Golgi apparatus which is then transferred to the immature secretory granule. If this has a more acidic interior, the protein can dissociate, allowing the receptor to recycle back to the Golgi apparatus for re-use. Such speculation about the existence of receptors for sorting proteins on specific pathways and the role of transmembrane pH gradients has received some support from the study of the synthesis and packaging of lysosomal enzymes in mammalian cells. After synthesis and translocation into the lumen of the endoplasmic reticulum, these are transferred to the Golgi apparatus, where a specific subset of high-mannose oligosaccharide side-chains are phosphorylated at the 6-carbon position of mannose. This alteration forms an address tag which allows these proteins to bind to an integral membrane mannose 6-phosphate (M6P) receptor protein. Membrane vesicles carry the lysosomal enzymes bound to M6P receptors from the Golgi apparatus to the lysosome (possibly via an endosomal compartment), where the more acid interior causes release of the enzymes from the M6P receptors, allowing recycling of the receptors to the Golgi.

Using triggered secretory pathways it is possible for there to be co-secretion of different proteins (and other secretory products) by the same cells. Detailed immunocytochemical studies on a variety of proteins secreted by exocrine pancreas have suggested that, in individual cells of this tissue, all secretory granules contain all the proteins to be secreted, in this case trypsinogen, chymotrypsinogen, deoxyribonuclease and ribonuclease. Unfortunately, immunocytochemistry is difficult to quantify and therefore it is not possible to assess accurately the possibility of microheterogeneity in terms of the relative amounts of different proteins in individual secretory granules.

The synthesis and intracellular packaging of insulin in pancreatic β cells

Insulin is made by, stored in and secreted from the β cells of the islets of Langerhans in the pancreas. The islets comprise only about 1–2 % of the weight of the pancreas, are innervated and have a good blood supply.

Individual islets can be isolated by partial collagenase digestion of whole pancreas and each consists of a cluster of approximately 5000 cells (Fig. 6.2). The islets of Langerhans perform several endocrine functions. The major cell type is the insulin-containing β cell, but glucagon-containing α cells and at least two other minor cell types producing other peptide

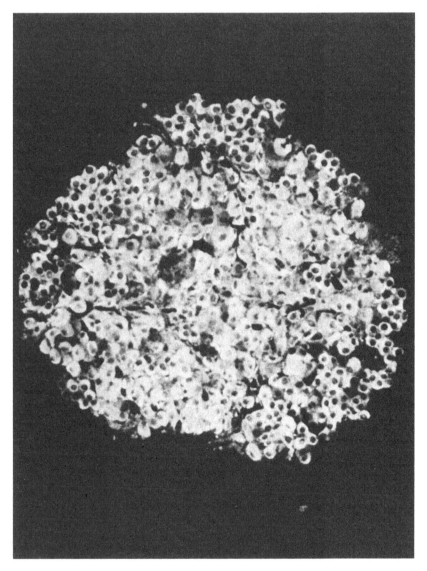

Fig. 6.2. Low-power, dark-ground photomicrograph of a rat islet of Langerhans isolated by collagenase digestion of rat pancreas. Micrograph kindly given by Miss Rosemary Jones.

hormones are also present. The synthesis and secretion of insulin show many of the basic characteristics of other triggered secretory pathways. Insulin is synthesized as a precursor protein and is stored in secretory granules (Fig. 6.3). Exocytosis is dependent on a stimulus and requires

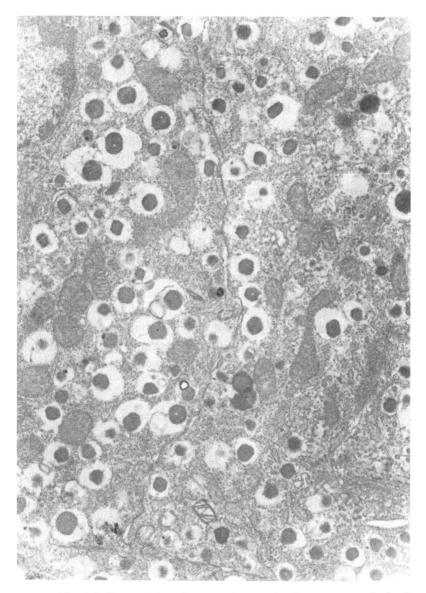

Fig. 6.3. Transmission electron micrograph of a rat pancreatic β cell showing many insulin granules. Micrograph kindly given by Prof. Lawrence Herman.

Ca^{2+} and energy. The intracellular network of events linking stimulus recognition to exocytosis is termed stimulus–secretion coupling. In experimental systems, exocytosis can be triggered when insulin biosynthesis is inhibited. There is much evidence supporting a role for intracellular messengers including Ca^{2+}, cyclic AMP and IP_3 in controlling the exocytotic process. Pharmacological disruption of the cytoskeleton interferes with stimulus–secretion coupling.

Insulin is a polypeptide hormone, the effects of which were described in Chapter 5. It consists of two peptide chains linked by disulphide bonds and is synthesized as a single-chain precursor with a signal peptide at the amino-terminus. This precursor is known as pre-proinsulin and, after cleavage of the signal peptide, gives proinsulin (Fig. 6.4), a hormonally inactive protein that passes from the endoplasmic reticulum to the Golgi apparatus and into secretory granules before proteolytic cleavage to give insulin. The existence of proinsulin was first inferred from studies of insulin biosynthesis in a human insulinoma which produced large amounts of insulin. When slices were incubated with ^{3}H-leucine, incorporation first occurred into a high molecular weight radioactively labelled polypeptide (proinsulin) and later into insulin, thus establishing the precursor–product relationship of proinsulin and insulin. The existence of proinsulin appears to be necessary for the correct folding and disulphide bonding of the insulin A and B chains. However, the subsequent discovery of many other stable prohormones even for single-chain peptides has suggested to some authors that their existence is more a reflection of the evolutionary origin of hormones, possibly as proteolytic breakdown products affecting membrane permeability to nutritional compounds.

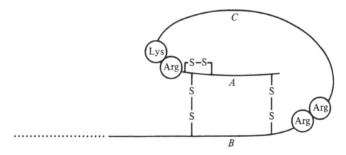

Fig. 6.4. Diagram of human proinsulin. The A chain (21 amino acids) and B chain (30 amino acids) are connected by the C-peptide (31 amino acids) through pairs of basic amino acids. The position of the signal peptide, cleaved during biosynthesis, is shown by the dotted line.

The maturation of proinsulin to insulin within the β cell and its passage from the endoplasmic reticulum to mature secretory granules has been studied using pulse-chase experiments and the techniques of subcellular fractionation and electron microscopic autoradiography. The overall pathway is summarized in Fig. 6.5, with the approximate time course indicated and data from pulse-chase experiments shown in Fig. 6.6. Proinsulin appears to be converted to insulin at a post-Golgi stage during maturation of the secretory granule. This maturation involves a change in morphology of the secretory granule, the new vesicle formed from the Golgi apparatus having a cytoplasmic protein coat which is lost during

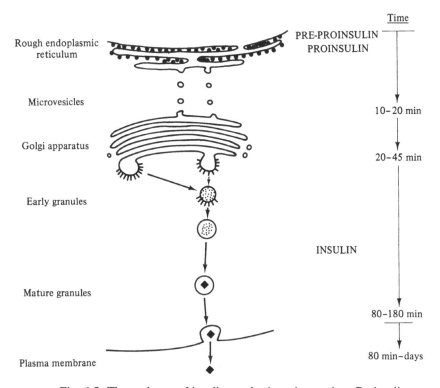

Fig. 6.5. The pathway of insulin synthesis and secretion. Proinsulin is packaged by the Golgi apparatus into secretory granules where it is progressively converted to insulin with a $t_\frac{1}{2}$ of approximately 1 h. The mature secretory granule contains in addition to insulin and *C*-peptide, Zn^{2+}, Ca^{2+}, ATP, amines and some other minor proteins which may be co-secreted, but whose physiological function is unknown. The eventual fate of the secretory granule may be fusion with the plasma membrane, allowing insulin secretion, followed by endocytic recovery of the membrane or autophagy by lysosomes.

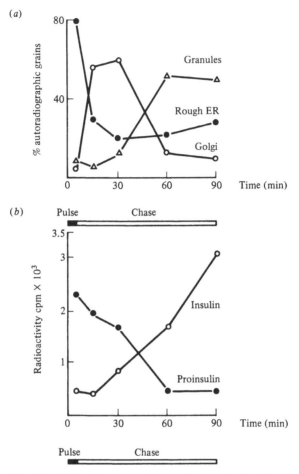

Fig. 6.6. Pulse-chase experiment demonstrating the passage of insulin and proinsulin through the subcellular compartments of cultured islet cells. A 5-min ³H-leucine pulse was followed by an 85-min chase period. Electron microscopic autoradiography showed the distribution of labelled polypeptides (*a*). Assessment of proinsulin and insulin content by chromatography of homogenates (*b*) showed that the time point at which the proinsulin and insulin curves cross corresponds to that at which the curve of radioactivity in the Golgi complex crosses with that in secretory granules. Data modified from Orci, L. (1985). *Diabetologia*, **28**, 528–46.

development (Fig. 6.7). Proteolytic enzymes with appropriate specificity for removing *C*-peptide from proinsulin by cleaving the molecule at the pairs of basic amino acids linking the *C*-peptide to the *A* and *B* chains have been localized in secretory granules. The effectiveness of this

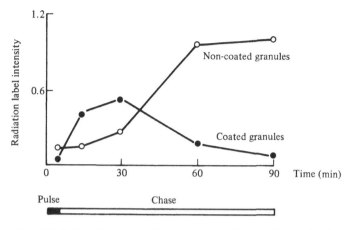

Fig. 6.7. Pulse-chase experiment showing the involvement of coated and non-coated granules in the maturation process of insulin secretory granules. Pulse-chase conditions were as described in Fig. 6.6. The morphology of granules was determined by electron microscopy and radiation label density by autoradiography. Data modified from Orci, L. (1985). *Diabetologia*, **28**, 528–46.

cleavage can be seen from the observation that, normally, 95 % of secreted protein is in the form of insulin and *C*-peptide, the remainder being proinsulin and partially cleaved proinsulin.

The mechanism of triggering insulin secretion

Investigation of the factors causing insulin release and the mechanism of stimulus–secretion coupling has been possible only since the development of immunoassays capable of measuring the low concentration of insulin found in the circulation or released in experimental systems. Immunoradiometric assays (Chapter 2) for insulin are sufficiently sensitive to be capable of detecting as little as 10^{-13}g of insulin, which is equivalent to approximately 1 % of that present in a single β cell. The availability of antibodies specific to *C*-peptide and monoclonal antibodies capable of distinguishing different partially cleaved forms of proinsulin has enabled investigation of the release of all of these from islets of Langerhans, and their subsequent fate in the circulation.

The major physiological stimulus of insulin secretion in Man is feeding, which results in a rise in serum concentration of blood glucose and amino acids. The glucose concentration giving half-maximum stimulation of insulin release is approx 12 mM, with a threshold at about 5 mM (Fig. 6.8). *In vivo*, a number of additional factors may modulate the response

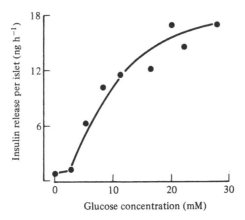

Fig. 6.8. The effect of glucose concentration on insulin secretion from isolated rat islets of Langerhans.

to glucose (Fig. 6.9). Glucose administered orally is more effective at causing insulin secretion than glucose administered intravenously, which suggests a stimulatory role for gut hormones. Stimulation of nerves to the pancreas may also affect insulin release and it has been proposed that activation of the parasympathetic and inhibition of the sympathetic nervous systems together help increase insulin secretion. Finally, a physiological role for the secretory products of other islet cells in controlling insulin secretion has not been excluded. Exogenously added glucagon can potentiate insulin secretion in isolated islets *in vitro*, but although this hormone is produced by α cells, it is secreted in response to a low blood glucose concentration, thus making it unlikely to be inolved in the physiological control of insulin secretion. It is of interest to note that glucagon is used therapeutically in the management of acute hypoglycaemia.

Investigation of insulin secretion from pancreatic islet cells *in vitro* has shown that the effect of exogenous agents is complex. Primary secretagogues such as glucose and some amino acids, e.g. leucine, stimulate secretion when present alone. Secondary secretagogues or potentiators have no effects on their own but alter the effects of primary secretagogues in one of two ways (Fig. 6.10). One group, characterized as metabolic fuels and including pyruvate and some amino acids, lower the glucose concentration threshold required for insulin secretion but have no effect on maximal secretion. The second group, including agents causing a rise in cyclic AMP concentration, e.g. theophylline (via inhibition of cyclic AMP phosphodiesterase) and glucagon, have no effect on the glucose

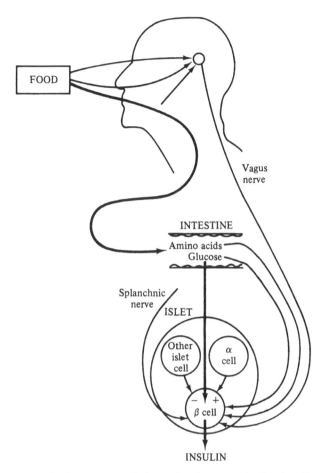

Fig. 6.9. A summary of the major factors affecting insulin secretion in Man. Reproduced from Hales, C. N. (1971). *Proc. Nutr. Soc.*, **30**, 282–8.

concentration threshold required for secretion but increase the maximum amount of insulin secreted in the presence of a high concentration of glucose.

There is still considerable doubt about the mechanism by which glucose acts as an insulin secretagogue, though two main hypotheses have been proposed. The 'regulator-site' hypothesis envisages a cell surface recognition system for glucose initiating the intracellular events leading to secretion. This model has clear analogies both with other triggered secretory systems, e.g. histamine release, and with the mechanism of action of cell surface binding hormones discussed in Chapter 5. The inability to identify a glucose-binding membrane receptor despite con-

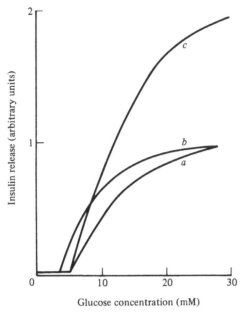

Fig. 6.10. The effects of secondary secretagogues (potentiators) on glucose stimulation of insulin secretion from pancreatic islets of Langerhans *in vitro*. The release of insulin is shown in the presence of (*a*) glucose alone, (*b*) glucose plus a potentiating concentration of pyruvate and (*c*) glucose plus a potentiating concentration of theophylline. Note that, in (*b*) , the threshold and half-maximal stimulatory concentrations of glucose are altered. In (*c*), only the maximum level of insulin secretion is affected.

siderable effort suggests that it is unlikely to be the correct mechanism for insulin secretion. The second hypothesis, known as the 'substrate-site' or 'fuel' hypothesis, proposes that glucose is metabolized within the β cell to generate a change in concentration of one or more key metabolic intermediates that act as an intracellular trigger for insulin release. The detailed examination of the effects of a large number of sugars, amino acids and some other compounds has shown an excellent agreement between the rates of metabolism, and the rates of insulin release as a function of concentration of the sugar or other compound. These studies, in addition to the effects of a number of metabolic inhibitors which prevent insulin secretion, have led to the suggestion that the key metabolic intermediate may be a phosphorylated intermediate or product of glycolysis (Fig. 6.11). There is no agreement over the identity of this metabolite.

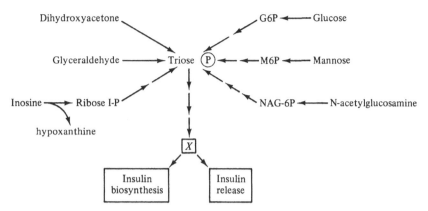

Fig. 6.11. The way in which metabolizable sugars and other compounds may stimulate insulin release and biosynthesis as a result of a rise in intracellular concentration of a metabolic (probably phosphorylated) intermediate X. Modified from Ashcroft, S. J. H. (1980). *Diabetologia*, **18**, 5–15.

Whichever of the above hypotheses of glucose stimulation of insulin secretion proves correct, there still needs to be a biochemical link to the secretory granules to trigger exocytosis. Dose-dependent glucose stimulation of intracellular accumulation of cyclic AMP and IP_3 prior to the initiation of insulin release have been reported. In addition, *in vitro* phosphorylation of granule membrane proteins has been observed. Thus, it is possible that cyclic AMP-dependent protein kinase and/or DAG stimulated protein kinase C-mediated phosphorylation may play a role in exocytosis as they do in many hormone-stimulated events. The rise in intracellular IP_3 concentration probably also releases Ca^{2+} from intracellular stores into the cytosol. A glucose-stimulated rise in cytosolic Ca^{2+} has been inferred from the results of $^{45}Ca^{2+}$ uptake and washout experiments and directly by measurements using Ca^{2+}-activated fluorescent indicators that can cross the plasma membrane. Through calmodulin, Ca^{2+} may activate many cell processes including, potentially, protein kinases. Thus, a consensus exists that cyclic AMP, IP_3 and cytosolic Ca^{2+} are all contributing as intracellular messengers for the amplification of expression of the glucose-induced secretory signal.

It has been known for many years that, in common with other exocytotic systems, removal of extracellular Ca^{2+} prevents glucose-stimulated insulin secretion in experimental systems. It is possible that a glucose-stimulated rise in cytosolic Ca^{2+} is at least partly due to an alteration of plasma membrane Ca^{2+} permeability rather than simply an effect on intracellular stores. Evidence for early changes in ionic fluxes across the plasma membrane includes the observation that glucose causes a depolarization of the β cell plasma membrane. Thus, glucose inhibition of K^+ permeability could trigger Ca^{2+} entry. Recent experiments have expanded these observations at a molecular level using the technique in which a small patch of plasma membrane is pulled from the cell surface on to a microelectrode and individual ion channels can be observed (patch clamp). In one laboratory, it has been shown that β cells have a plasma membrane K^+ channel sensitive to glucose metabolism, and, in another, that there is a K^+ channel inhibited by low concentrations of ATP. The further development of these observations offers the hope of a complete description of the plasma membrane ionic events associated with glucose stimulation of insulin secretion.

Analysis of the time course of glucose stimulation of insulin secretion from perfused isolated pancreas (Fig. 6.12) suggests that the triggering of secretion may have components other than simply a rise in intracellular messenger concentration. The selective effects of inhibitors on the two phases of the time course has suggested that the initial release of insulin may be from secretory granules close to the plasma membrane, the second, prolonged phase being from deeper lying and newly synthesized granules. It is likely that components of the cytoskeleton play a role in secretion, microtubules helping to route secretory granules to the plasma membrane and a microfilamentous web being involved in their localization close to the membrane. Agents disrupting normal microtubule and

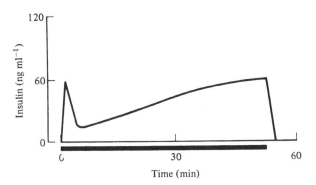

Fig. 6.12. The effect of prolonged glucose infusion on insulin secretion from perfused rat pancreas. The bar shows a period of stimulation with 16 mM glucose. Data selected from Grodsky, G. M. *et al.* (1970). In *Structure and Metabolism of the Pancreatic Islets*, pp. 409–20, (edited by S. Falkmer *et al.*) Pergamon Press: Oxford.

microfilament structure will interfere with normal insulin secretion in experimental systems.

While it is possible to inhibit protein synthesis in isolated pancreas without preventing the initial effect of glucose to stimulate insulin release, glucose stimulation of insulin synthesis normally accompanies the increased insulin secretion. The main intracellular messenger involved may be the phosphorylated intermediate or product of glucose metabolism thought to trigger secretion. The insulin gene has been cloned and there is currently considerable interest in possible regulating nucleic acid sequences upstream of the encoding sequences. It appears that glucose stimulation of synthesis may involve effects at both the transcriptional and translational level. There may be a common mechanism triggering the synthesis of other components on the secretory pathway, since during secretion there is also an increase in synthesis of granule membrane proteins.

The origin of diabetes

Diabetes mellitus is now thought to be a heterogeneous group of diseases arising from a complex interaction between the genetic constitution of the individual and specific environmental factors. Diabetes is a common condition affecting 1–2 % of many populations. In the USA, the figure is approaching 5 %. While the acute and often lethal symptoms of the disease can be controlled, the long-term complications which include blindness, kidney disease, gangrene and possibly heart disease reduce life expectancy by up to a third.

Broadly, patients can be classified as having Type I (sometimes known as juvenile-onset) or Type II diabetes (Table 6.1). In Type I diabetes (which is less common), there is an absence of insulin, preventing glucose uptake into the peripheral tissues, increasing protein breakdown and lipolysis. Fatty acids released by adipose tissue are converted to ketone bodies (acetoacetate, β-hydroxybutyrate) in the liver leading to ketoacidosis. The disease is fatal without insulin therapy and it was a major breakthrough when pancreatic insulin extracts became available in the 1920s and immediately kept patients alive who would otherwise have died. Type II diabetics may have an elevated fasting blood glucose concentration but have normal or high insulin levels. It is sometimes difficult to diagnose a Type II diabetic without the aid of a glucose tolerance test (Fig. 6.13), when glucose is cleared less rapidly from the blood than in a normal individual. This, together with the normal circulating levels of insulin observed, suggests that Type II diabetes may result from insulin resistance of the peripheral tissues. Alternatively, it

Table 6.1 *Typical presentation of patients with diabetes mellitus*

Type I	*Boy, aged 7. Grandfather a diabetic on insulin*
	2–3 week history of thirst, polyuria and weight loss. Increasingly drowsy over the last 2–3 days.
	On examination, found to be semi-comatose, rapid respiration, breath smells fruity, loss of skin elasticity. Abscess on right calf. Urine dip-stick tests: glucose +++, ketones ++. Plasma glucose concentration 25 mMolar (normal is 5 mM).
Type II	*Woman, aged 55* Attended optician due to failing vision last 2 years. Referred to GP by optician,? diabetes.
	On examination, very obese, slightly depressed reflexes, otherwise no abnormality detected. Eye examination: cataract both eyes, retina where visible shows haemorrhages and exudates. Urine dip-stick test: glucose negative; ketones negative Plasma glucose concentration 12 mM

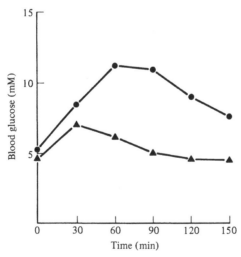

Fig. 6.13. The glucose tolerance test. After an overnight fast, the patient takes by mouth a standard dose of 50 g glucose dissolved in 100 ml water. Blood glucose is measured at intervals over the subsequent 3 h. Typical glucose tolerance curves for a normal person (▲) and a mild diabetic (●) are shown.

may be the result of an inappropriate secretory response to blood glucose in terms of either the magnitude of insulinotropic response or its dynamics. Type II diabetics are usually mature adults and overweight. The obesity may be the cause of the disease and initial treatment is dieting. Type II diabetics may also require treatment with oral hypogly-caemic agents (such as sulphonyl ureas) which potentiate glucose-induced insulin release from β cells, and some require insulin therapy. Some Type II diabetics are treated initially by dieting, later requiring sulphonyl ureas and finally insulin therapy.

While the origins of diabetes remain uncertain, a number of clues have been found. Studies on identical twins have suggested that, in Type II diabetes, genetic factors are predominant. For Type I diabetes, emphasis has been placed on finding an agent that can rapidly cause degranulation of β cells leading to the acute disease. Viruses causing diabetes in mice have been isolated, but there is no single virus that has been proven to cause diabetes in Man. Studies on activation of the immune system in the early phases of Type I diabetes have provided some insight into the possible pathology. Complement-fixing, circulating antibodies to β cell membrane components appear to be a good indicator of the Type I disease, as do the appearance of some lymphocyte subsets in the pancreas. It is tempting to speculate that autoimmune events (or immune responses to viral infection) causing membrane damage to the β cell play a role in the development of the disease.

7

The genome

Genes and disease

The human genome codes for approximately 50 000 structural genes containing information which is essentially expressed in the structure and biological functions of proteins. Approximately 1400 of these genes are known (when 'defective') to lead to a recognizable human disease. In about one-third of these 'single-gene' disorders, the chromosomal localization of the gene is known, approximately two-thirds of these localizations are on autosomes (the 22 pairs of non-sex chromosomes), the remainder on the X-chromosome. Of the diseases known to be attributable to a single 'defective' gene, the biochemical mechanism of the defect has been worked out in about 250 cases. As a generalization, 'single-gene' diseases which are inherited as autosomal dominant disorders (about 600 known) do not produce easily decipherable biochemical abnormalities and the detailed mechanism of the disorder is known in only a handful of diseases in this category. Autosomal recessive disorders, however, frequently lead to the defective functioning of a single enzyme in a metabolic pathway, and the majority of the approx. 200 single enzyme defects which are known have been discovered in this disease group. Single-gene disorders are relatively rare and, if all are grouped together, the incidence is barely 1 in 100 (Table 7.1). However, attempts to estimate the total load of genetic disease including common disorders with a strong genetic component (Table 7.2) and congenital malformations raise the incidence about fivefold. Even this figure underestimates the importance of genetic disease, since it accounts for up to 30 % of paediatric admissions to hospital and is an important cause of death in young people.

In the past decade, the development of 'recombinant DNA technology' has led to remarkable advances in techniques available to pinpoint genetic faults leading to disease. Since the early and mid 1970s, it has been possible to 'tailor' specific pieces of DNA, and insert them into

Table 7.1 *The total load of genetic disease*
The bracketed figures are approximations and include the genetic
contribution, being one-third of disorders such as schizophrenia, diabetes
mellitus and epilepsy and half of malformations such as spina bifida and
congenital heart disease.

Type of genetic disease	Frequency per 1000 population
Single gene	
Dominant	1.8–9.5
Recessive	2.2–2.5
X-linked	0.5–2.0
Chromosome abnormalities	6.8
Common disorders with a significant genetic component	(7–10)
Congenital malformations	(19–22)
Total (approximate)	(37.3–52.8)

Data from Weatherall, D. J. (1985). *The New Genetics and Clinical Practice*
(2nd edn). Oxford University Press.

Table 7.2 *Common disorders with a strong genetic component*

Disorder	Frequency per 1000 population
Schizophrenia	8–10
Manic-depressive disorders	4
Epilepsy	5
Diabetes mellitus	3–10
Total	20–29

Data source as for Table 7.1.

'vectors' which can carry them into bacteria (or even into eukaryotic
cells) where the specific piece of DNA can be reproduced many times
over or 'cloned'. It is also now possible, by *in vitro* copying procedures, to
sequence accurately pieces of DNA many thousands of bases long and
hence to span the sequence of entire genes coding for proteins, including
those regions which exert regulating functions. Such methods have not
only led to an increasing number of specific gene probes for use in
detecting genetic abnormalities, but also resulted in the ability to
sequence sections of DNA (and hence predict the primary amino acid
sequence of rare protein molecules) and the production of large amounts
of therapeutically useful proteins in bacterial cultures.

Genes and the sites of genetic disease can now be localized to specific
chromosomes using cells containing mixtures of mouse and human

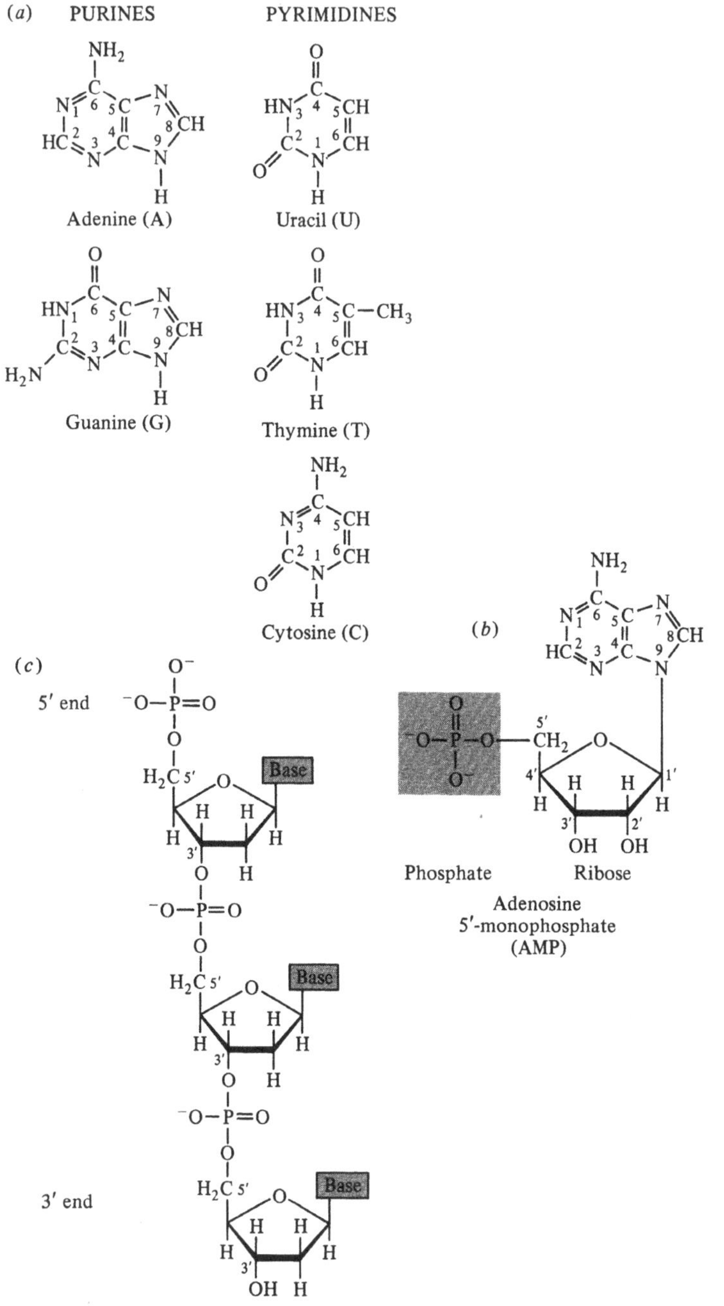

Fig. 7.1. (*a*) The common bases occurring in DNA and RNA. Note that uracil (U) replaces thymine (T) in RNA compared to DNA and that spontaneous deamination of cytosine (C) converts it to uracil.

chromosomes (somatic cell hybrids), by isolating individual chromosomes by fluorescence-activated sorting (p. 48), by *in situ* localization using radioactive nucleic acid probes (p. 240), and by techniques of 'chromosome walking'. These, along with the ability to separate and study very large fragments (several hundred thousand bases long) of DNA by techniques such as pulsed-field electrophoresis, are now making it possible to localize genes responsible for many diseases to specific chromosomes and potentially to sequence the whole human genome (3.5×10^9 base pairs) itself. Medical texts and journals often summarize these recent advances as the 'new genetics', a term first used in 1979 as the potential impact on human disease became apparent.

Recombinant DNA techniques have also led to the prospect of gene therapy. Thus, experiments have been conducted where foreign DNA has been introduced into the nuclei of animal eggs, resulting in the production of offspring with the inserted DNA in terminally differentiated cell types. To date, such transgenic mice, rabbits, sheep, pigs and even fish have been produced and there are obvious implications for the treatment of inherited disease in Man.

This chapter reviews the structure of and means of expressing the genetic material and discusses the basic approaches required for recombinant DNA technology. The methods of using gene probes and possibilities of gene therapy are then described, and finally there is discussion of oncogenes, a fuller knowledge of which may lead to much better treatment and perhaps even the eradication of cancer.

DNA, RNA and the genetic code

The four bases in DNA are adenine (A), guanine (G), cytosine (C) and thymine (T). These are attached to a repeating sugar–phosphate backbone as shown in Fig. 7.1. A and G are purine bases; C and T are pyrimidine bases. The sugar in DNA is desoxyribose, hence desoxyribonucleic acid; in RNA, the sugar is ribose, hence ribonucleic acid. In RNA, thymine is replaced by a different pyrimidine base, uracil (U). The carbon and nitrogen atoms in purine and pyrimidine bases are numbered

Fig. 7.1. (*cont.*)

(*b*) A purine base, e.g. adenine (A), becomes a nucleoside (e.g. adenosine) when covalently joined to the sugar desoxyribose or ribose, and a nucleotide (adenosine 5′-mono-, or di- or tri, phosphate, i.e. AMP, ADP or ATP) when phosphorylated on the 5′-position of the sugar residue. (*c*) Nucleotides in DNA and RNA are polymerized by 5′→3′ phosphodiester linkages as shown. Each nucleic acid chain therefore has a free 5′ and a free 3′ end.

as shown in Fig. 7.1, in each case a nitrogen atom (N9 in purines, N1 in pyrimidines) is bonded to the 1 carbon of the pentose sugar ring. The atoms of the pentose sugar are labelled 1' to 5' (the 'prime' sign indicating the positions on the pentose sugar ring rather than the carbon atoms of the base), and ribose (in RNA) differs from desoxyribose (in DNA) by having a hydroxyl (OH) group on the 2' carbon atom rather than a hydrogen (H) atom. Ribose and desoxyribose are covalently linked by phosphodiester bonds between the 3'-OH of one sugar and the 5'-OH of the adjacent sugar, as shown in Fig. 7.1(c). Therefore, one end of the sugar–phosphate backbone of the nucleic acid chain (the 5' end) has a free 5'-OH and the other end (the 3' end) has a free 3'-OH. DNA normally exists as a double helix (the Watson–Crick double helix, Fig. 7.2) such that A pairs with T and G pairs with C, one strand of the helix running in a 5' to 3' direction, the other in an antiparallel 3' to 5' direction. Three hydrogen bonds are possible between G–C base pairs and two between A–T base pairs.

Theoretically, a right-handed double helix with A–T and G–C base pairing is the most stable structure. While G–T or C–T base pairing would be possible with a minor distortion of the helix, these have not been found in DNA. Some DNA crystals have been shown to have a left-handed double helix, termed Z-DNA. It is not clear whether areas of Z-DNA occur *in vivo* and whether such regions act as signals in the overall DNA structure. Because of complementary base pairing, the sequence of bases on each DNA strand can be copied exactly into new strands, to be passed on to daughter cells. This is termed replication and is performed by an enzyme called DNA polymerase. Again because of complementary base pairing, the sequence of bases in one DNA strand can be exactly copied into an RNA strand (with U replacing T and base pairing with A). This is done by RNA polymerase and is termed transcription. RNA containing information for protein synthesis passes from the cell nucleus to cytoplasmic ribosomes, where the specified nucleic acid sequence is copied into an amino acid sequence. This is termed translation. The majority of the RNA in a cell is considered to be single-stranded, but can form internal double-helical regions by folding back upon itself, i.e. an RNA molecule can have a precise secondary structure. Most of it (85 %) is ribosomal RNA (rRNA), performing an important structural role in the cytoplasmic ribosome, and most of the rest (appoximately 13 %) is transfer RNA (tRNA), which acts as an 'adaptor' in the translation process converting the sequence of bases in RNA into an amino acid sequence. Only a small proportion (about 1 %) of total cellular RNA acts as messenger RNA (mRNA) carrying information from the cell nucleus to the cytoplasm. This small proportion of RNA is itself derived from

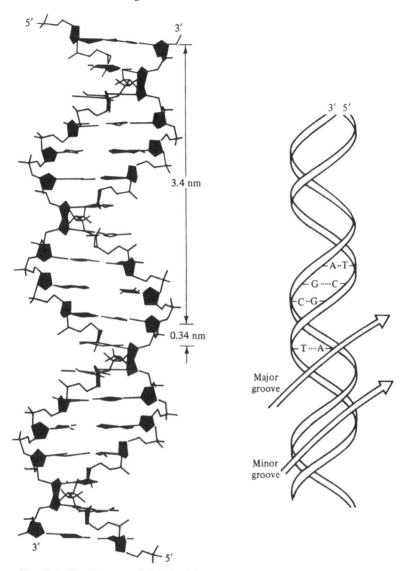

Fig. 7.2. The Watson–Crick double helix. The right-handed helix consists of two anti-parallel intertwined strands held together by hydrogen-bonded base-pairing between A and T, and C and G. The exterior of the helix shows a major and minor groove; in the interior of the helix base pairs are stacked 0.34 nm apart. The helix makes a complete turn approximately every 10 base-pairs. The helix is less rigid than the α-helix found in proteins because there are no hydrogen bonds between successive segments. One advantage of using DNA as the genetic material is that the order of bases along the interior of the helix (i.e. the specification of information) can be varied without altering the secondary structure of the whole molecule.

larger RNA molecules (heterogeneous nuclear or hnRNA) which are the primary products of the transcription of structural genes. DNA and all forms of RNA are complexed with proteins within the cell (as nucleoprotein complexes), a condition which has various structural and functional consequences.

The assembly of DNA, RNA and protein chains within a cell is a stepwise process proceeding in a 5' to 3' direction (or from the amino to the carboxyl terminus in the case of a polypeptide chain). One strand of a DNA double helix (the information or 'sense' strand as opposed to the 'nonsense' strand) is copied into RNA, with the sequence of bases containing information for protein synthesis in a 'triplet code', i.e. three bases code for one amino acid. This genetic code is shown in Table 7.3. Each triplet of bases is called a codon. Note that the genetic code, as originally deciphered, is an RNA code rather than a DNA code, as U replaces T, i.e. the order of bases shown is the order found in mRNA. The linear array of bases in mRNA, grouped into threes, precisely specifies the order of amino acids to be sequentially added to the growing polypeptide chain. All polypeptide chains start with methionine (the 'initiator' codon AUG, rarely GUG), which is removed post-translationally in most proteins. Three codons – UAA, UAG and UGA – signal 'stop', i.e. polypeptide chain synthesis halts when one of these 'termination' codons is reached. There are 64 possible ways to arrange four bases into triplet codons (Table 7.3). As there are only 20 amino acids in proteins, some triplets code for more than one amino acid, e.g. leucine, serine and arginine have as many as six codes each, while tryptophan has only one. The code is therefore said to be 'redundant', i.e. each amino acid can have more than one codon, but it is also 'unambiguous', i.e. each triplet specifies only one amino acid. Note that there are no 'commas' in the code and also that there are no 'nonsense' codons, i.e. none of the 64 possible codons has an unassigned function. Any sequence of DNA can therefore be 'read' as an amino acid sequence in several different 'frames' (each specifying a different amino acid sequence) provided that the frame is 'open', i.e. provided that it is not interrupted by one of the three termination codons, UAA, UAG or UGA. Note also that the code is 'universal', i.e. with minor exceptions in some codons in the DNA of mitochondria and ciliate protozoans, the same codon specifies the same amino acid in all organisms.

The flow of information from DNA to RNA to protein is termed the 'central dogma'. Since this was first enunciated, enzymes have been discovered, particularly in viruses, which can copy RNA into DNA. These are called the reverse transcriptases. There is, however, as yet no

Table 7.3 *The genetic code*
This is shown in its RNA form with U replacing T. The code is commaless, redundant, unambiguous and more-or-less universal.

First position (5' end)	Second position				Third position (3' end)
	U	C	A	G	
U	Phe	Ser	Tyr	Cys	U
	Phe	Ser	Tyr	Cys	C
	Leu	Ser	Stop (och)	Stop	A
	Leu	Ser	Stop (amb)	Trp	G
C	Leu	Pro	His	Arg	U
	Leu	Pro	His	Arg	C
	Leu	Pro	Gln	Arg	A
	Leu	Pro	Gln	Arg	G
A	Ile	Thr	Asn	Ser	U
	Ile	Thr	Asn	Ser	C
	Ile	Thr	Lys	Arg	A
	Met	Thr	Lys	Arg	G
G	Val	Ala	Asp	Gly	U
	Val	Ala	Asp	Gly	C
	Val	Ala	Glu	Gly	A
	Val (Met)	Ala	Glu	Gly	G

known way in which information from DNA, once it has been transferred to protein, can be returned to the genome. The biochemical differences between the two nucleic acids are that U replaces T and that the hydroxyl group on the 2' position of the pentose sugar ring is missing in RNA compared to DNA. The absence of the hydroxyl group in DNA makes the molecule chemically less labile and its presence in RNA may act as a recognition site for enzymes to break down RNA. DNA is designed as the stable genomic material; different species of RNA need to be 'unstable' and capable of being broken down during the cell cycle. The presence of thymine rather than uracil in DNA may be due to the spontaneous deamination of cytosine in both DNA and RNA which converts this base to uracil (see Fig. 7.1). Enzymes are present in the nucleus which can convert uracil back to cytosine, i.e. the presence of U in DNA is recognized as a mistake (since a G–C pair is turned into a G–U which on replication could result in an A–U pair) and corrected. If U was used as a pairing base in DNA, those derived from C could not be distinguished from those originating as U and the spontaneous deamination of C could not be corrected. The presence of an incorrect U in RNA, on the other hand, cannot be recognized and corrected. The presence of T rather than U in DNA again helps stability.

The structure of the human genome

It is estimated that the human genome contains approximately 3.5×10^9 base pairs of DNA distributed unevenly between 46 chromosomes. (In all diploid cells, there are 23 chromosome pairs, numbered 1 to 22 in decreasing order of size, plus X and Y sex chromosomes.) The total length of DNA in a human cell (measured as a Watson–Crick double helix) is approximately 2 m. This is packaged into a cell nucleus typically $5 \mu m$ across, involving a 'packaging ratio' of at least 1:10 000. While individual chromosomes are not visible, the DNA and protein complex derived from them is seen as the familiar chromatin of the interphase nucleus. This can still be highly condensed to form heterochromatin or show a more dispersed form as euchromatin (Fig. 7.3). Different areas of a typical cell nucleus can show different proportions of heterochromatin and euchromatin. As a generalization, heterochromatin is considered to be relatively 'inactive' and the more extended euchromatin is considered to be 'active' in gene transcription.

The first stage in the packaging of a DNA double helix into a more condensed structure is into a nucleosome filament, a structure with a diameter of about 10 nm. For many years it was known that eukaryotic nuclei contained a group of highly conserved basic proteins (i.e. with a high content of arginine and lysine) which were present in amounts equal to the DNA content of the cell. These histone proteins (molecular weight 11–21 kDa) were of five types – H1, H2A, H2B, H3, H4 – with some specialized cells such as chicken erythrocytes having a different histone (H5) replacing H1. Experiments in the early 1970s established that a repeating unit (a nucleosome) existed in interphase chromatin. This consisted of an octamer of histones (two each of H2A, H2B, H3 and H4) with 200 base pairs (bp) of DNA wound around the outside such that the array of histone octamers complexed with the wound double-helical DNA appeared as a 'string of beads' under the electron microscope. Of the 200 bp in a typical nucleosome, about 140 are closely complexed with the histone octamer; the rest is 'linker' DNA passing from one nucleosome to the next. The fifth histone (H1) is associated with this linker DNA (Fig. 7.4). During DNA replication, newly synthesized histone octamers preferentially associate with newly synthesized DNA, and individual histones do not appear to interchange between old and new

(a)

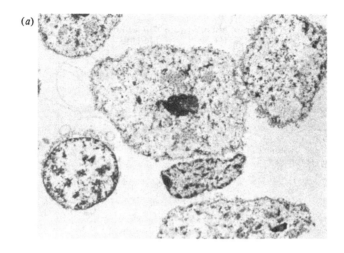

(b)

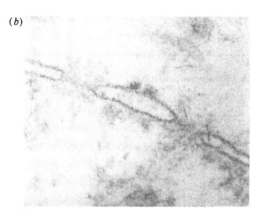

Fig. 7.3. Chromatin in cell nuclei. (*a*) Cell nuclei which have been isolated by homogenization and sub-cellular fractionation and examined by transmission electron microscopy are shown. The large nuclei with prominent nucleoli (sites of ribosomal RNA synthesis) contain almost entirely 'active' euchromatin; the smaller nuclei have areas of dark, condensed, 'inactive' heterochromatin. (*b*) A higher magnification of nuclear pore structures believed to be the routes of nucleocytoplasmic exchange.

histone octamers. Most of the DNA in a eukaryotic nucleus is probably organized into a nucleosome filament. While nearly all cells in the body, e.g. liver or kidney cells, have 200 bp of DNA per nucleosome, large neurones in the cerebral cortex (apparently uniquely among mammalian cells) have only 160 bp. Some specialized cells such as sea urchin sperm

(a)

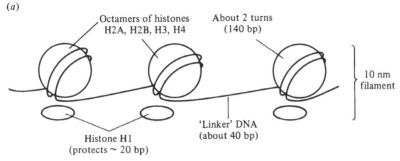

Octamers of histones
H2A, H2B, H3, H4

About 2 turns
(140 bp)

10 nm
filament

Histone H1
(protects ~ 20 bp)

'Linker' DNA
(about 40 bp)

(b)

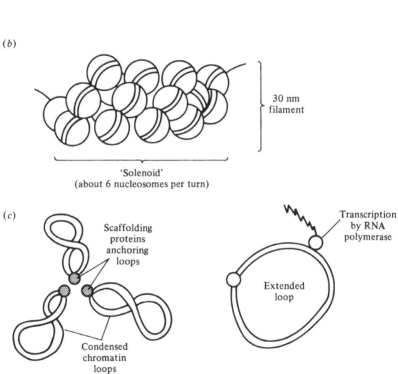

30 nm
filament

'Solenoid'
(about 6 nucleosomes per turn)

(c)

Scaffolding
proteins
anchoring
loops

Transcription
by RNA
polymerase

Extended
loop

Condensed
chromatin
loops

Fig. 7.4. Nucleosomes, solenoids and chromatin loops. (*a*) Double-helical DNA is wound around octamers of histones as the basic repeating unit in chromatin structure. (*b*) The nucleosome filament is further wound into a solenoidal structure. In chromosomes and interphase nuclei, this higher-order filament is condensed into loops of chromatin which are 'organized' by 'scaffolding' proteins. (*c*) Active transcription of genes requires the local dismantling of this complex chromatin architecture to allow access of the components of the transcription machinery.

have 240 bp per nucleosome. The significance of different chromatin repeat lengths between different cells is not yet clear. It is also possible that local areas of the nucleus may contain different nucleosome spacings from that of the bulk of DNA.

The next higher-order folding of the 10 nm nucleosome filament is into the 30 nm filament which can be seen in extended chromatin in the interphase nucleus. This structure appears to be a 'solenoid' with six nucleosomes per turn and possibly with the linker DNA occupying the centre of the coil. The folding of the nucleosome filament into the compact solenoidal 30 nm filament requires histone H1 (Fig. 7.4(*b*)). Current views on the further packaging of the 30 nm filament into condensed interphase chromatin and into metaphase chromosomes invoke a considerable degree of 'architecture' within the interphase nucleus. In recent years, much evidence has accumulated for the existence of a nuclear or chromosomal 'scaffold' (or matrix) composed of proteins which can bind areas of 'looped' DNA formed from further higher-order twisting and coiling of the 30 nm filament described above.

The nuclear envelope consists of an inner and outer nuclear membrane which fuse at the nuclear pores. These are complex structures thought to be involved in nuclear–cytoplasmic interchange and possibly in nuclear RNA processing en route to the cytoplasm (Fig. 7.3). The nuclear lamina is a structure between the nuclear envelope and the chromatin and is postulated to form a lattice which acts as an architectural framework for the nuclear envelope and as an anchoring point for interphase chromatin. The nuclear lamina can sometimes be seen by electron microscopy directly beneath the nuclear envelope; in other cells, it can be demonstrated only by isolating and extracting nuclear envelopes. The lamina is largely formed by three proteins termed lamins A, B and C. These proteins show striking sequence homology to intermediate filament proteins and show the same structural motif of a central α-helical rod portion with head and tail pieces attached (see p. 62). Nuclear lamins appear to belong to the same multi-gene family as intermediate filament proteins.

Isolation of metaphase chromosomes and extraction with acid to remove histones reveals a core or 'scaffold' running along the axis of the chromosome with loops radiating out from the core, each loop containing 30–100 kbp of DNA. This is shown diagrammatically in Fig. 7.4. Analysis of the proteins forming the chromosomal scaffold and acting as 'loop-fasteners' has shown two major proteins (SC 1 and SC 2), one of which is now known to be identical to topoisomerase II enzyme. Topoisomerases are enzymes which 'manage' the problem of superhelicity in higher-order

DNA structures. Superhelicity was originally discovered in closed circular DNA from animal viruses which form supercoils or supertwists, i.e. show superhelicity (Fig. 7.5). Closed circular DNA, or linear DNA with fixed ends as with chromatin loops in metaphase chromosomes, 'redistributes' local structural strain (e.g. produced by strand separation for purposes of replication or transcription) over long distances and produces superhelical structures. This behaviour, which occurs with naked DNA, is an even more complicated phenomenon in large DNA protein complexes and requires complex mathematical description. Topoisomerases can 'relax' strain in large supercoiled structures by cutting DNA strands and allowing 'knots' to unwind. Topoisomerase II is capable of cutting one double helix, allowing a different part of the strand

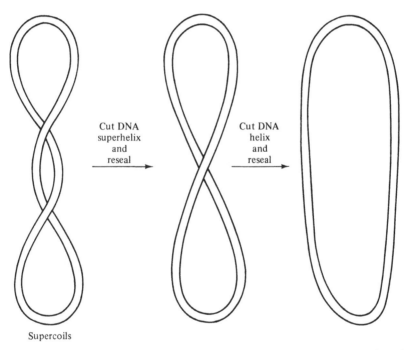

Cut DNA
superhelix
and
reseal

Cut DNA
helix
and
reseal

Supercoils

Fig. 7.5. Superhelicity in DNA. Long closed loops of DNA (e.g. in a virus or as long chromatin loops in eukaryotic nuclei) can adopt superhelical forms as a way of redistributing 'strain' within the closed structure. Topoisomerases can relieve this by breaking double-stranded DNA and resealing. Such enzymes are involved in 'managing' large chromatin loops inside the nucleus. Clinically they are important because they appear to be targets for cytotoxic drugs used particularly in the treatment of breast cancer. Topoisomerases are also implicated as antigens in some 'autoimmune' human disease such as scleroderma.

to pass through the cut, and then resealing the original incision. The existence of such an enzyme activity as a major component of the chromosome scaffold has obvious functional implications in the orderly folding and packaging of large chromatin loops.

While a relatively simple protein scaffold can be demonstrated in condensed metaphase chromosomes, decondensation to interphase chromatin requires a considerably more complex internal nuclear protein scaffold. Topoisomerase II, however, is known to be present in the nuclear matrix of interphase nuclei. A current (though speculative) concept of the way chromatin loops might be organized in the interphase nucleus is shown in Fig. 7.4.

Transformation

The property of DNA which provided the most conclusive evidence that it carries genetic information was its ability to transform recipient cells. This was originally shown with strains of *Pneumococcus*, the bacterium causing pneumonia in Man and other susceptible mammals. In the 1920s, Griffith studied a mutant form of the bacterium called an *R* form which was non-pathogenic to mice and grew as rough colonies on agar plates. This contrasted with the pathogenic *S* form which grew as smooth colonies on agar plates. When heat-killed *S* bacteria and live *R* bacteria were mixed either by injecting into mice or *in vitro*, live *S* bacteria which were pathogenic and grew as smooth colonies on plates were formed. This transformation of the *R* form to the *S* form could also occur when *R* form cells were mixed with a cell-free extract of *S* bacteria. Identification of the transforming factor as DNA came from careful chemical extraction and fractionation of cell extracts by Avery and collaborators in the 1940s.

Although different cells show differing susceptibility to transformation by DNA, it is a universal phenomenon. Thus, an hereditary trait lost through mutation can be restored by treating mutant cells with DNA isolated from wild type and, conversely, wild-type cells can be transformed to a mutant by treatment with mutant DNA. The latter is easily shown where the mutant type can be positive selected, as in the case of mutation to antibiotic resistance. In practice, the efficiency of transformation is usually very low (often less than 1 in 10^6 cells) but may be improved by handling cells in specific ways. Thus, pretreatment of *Escherichia coli* with $CaCl_2$ renders them receptive, and cultured animal cells can assimilate DNA supplied as a co-precipitate with calcium phosphate. The mechanisms by which DNA can cross the plasma membrane and become integrated into the genome are not well understood. Nonetheless, the practical ability to transform cells simply by adding

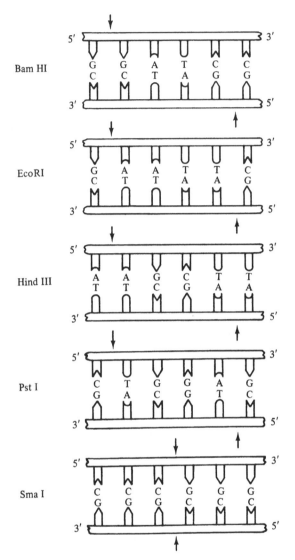

Fig. 7.6. The action of restriction enzymes. The restriction enzyme Bam HI (named because it is from the bacterium *Bacillus amyloliquefaciens H*) recognizes the sequence of six base pairs shown and cuts where indicated (↓) to create staggered ends. The enzymes Eco RI (from *E. coli RY13*) Hind III (from *Haemophilus influenza RD*) and Pst I (from *Providencia stuartii*) work in a similar way. The enzyme Sma I (from *Serratia marcescens*) also recognizes a sequence of six base pairs but produces a flush cut with blunt ends. For the analysis of DNA, restriction enzymes recognizing sequences of six base pairs are the most useful because they cut on average only once in every 4000 base pairs.

exogenous DNA has been essential to the development of gene cloning and allows amplification of any piece of DNA which transforms cells simply because of cell division and the replication of this DNA.

Cutting and splicing DNA

The ability to analyse and manipulate DNA was revolutionized in the 1970s by the discovery of restriction enzymes in bacteria. These enzymes recognize specific sequences in DNA and act to protect bacteria from intracellular foreign DNA, such as that deriving from bacteriophage attack. Host DNA is protected, even if suitable base sequences are present, by strategies such as base methylation which prevent enzyme attack. For experimental purposes, the most useful restriction enzymes are the type II restriction endonucleases of which several hundred have been characterized. The vast majority recognize and break DNA within specific sequences of tetra-, penta-, hexa- or hepta-nucleotides which have an axis of rotational symmetry (Fig. 7.6), sometimes described as palindromic. In some cases, the enzymes produce a flush cut and in others a staggered cut with protruding ends.

Since restriction enzymes recognize specific nucleotide sequences, it is possible to break up any DNA into specific restriction fragments which can then be separated according to size by gel electrophoresis (Fig. 7.7). Thus, it is possible to obtain physical maps of genes (Fig. 7.8), to identify specific viruses or parasite strains, to check for the presence of a certain gene or sequence in a specific cell or chromosome and to obtain specific DNA fragments which can be sequenced, used for genetic manipulation or tested for function (e.g. for protein binding or transfer of genetic information).

It is possible to incorporate restriction fragments into other pieces of DNA to create new recombinant DNA molecules. This is particularly simple when fragments with staggered ends have been produced since they will be attracted to the reciprocal staggered ends (or sticky ends) of another piece of DNA cut by the same restriction enzyme. The joined pieces of DNA can then be annealed by another enzyme, DNA ligase. DNA ligases capable of joining blunt-ended restriction fragments have also been isolated, though this is usually a less efficient process than with overlapping sticky ends. The ability to achieve blunt-end ligation is also important for creating recombinant DNA molecules from DNA fragments obtained by non-specific techniques such as mechanical shearing. Recombinant DNA molecules prepared by such techniques are essential when it is necessary to amplify a piece of DNA into many identical copies by cloning.

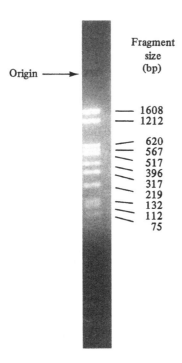

Origin ⟶

Fragment
size
(bp)

—— 1608
—— 1212

—— 620
—— 567
—— 517
—— 396
—— 317
—— 219
—— 132
—— 112
75

Fig. 7.7. Gel electrophoresis of nucleic acids. At neutral pH, the charge on nucleic acids is carried solely by the phosphate groups. All types of nucleic acid will thus have the same charge per unit mass regardless of base composition or length, and can therefore be separated by gel electrophoresis on the basis of size. Agarose or polyacrylamide gels are used. After electrophoresis, the separated bands are visualized by staining with ethidium bromide and illuminating with ultraviolet light. Orange bands on a dark background are seen. As little as 50 ng DNA in one band can be detected, and resolution can be such as to separate nucleic acids differing in length by 0.5 % (i.e. a fragment 200 bases long from one of 201 bases). Just as for protein gel electrophoresis in the presence of SDS (Chapter 1), the distance migrated in the gel is proportional to the logarithm of the M_r. Standards of nucleic acid fragments of known size are run as markers. The example shown is electrophoresis of Hind III restriction fragments of a plasmid in a 2 % agarose gel.

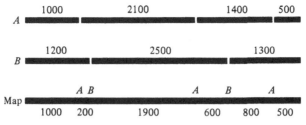

Fig. 7.8. Restriction mapping. A restriction map is created by incubating aliquots of DNA separately and together with different restriction enzymes (here called *A* and *B*) and identifying the size of fragments produced by electrophoresis. The map is thus a linear sequence of sites separated by defined distances (measured in base pairs) on the DNA molecule.

Techniques of DNA cloning: how to make a library
In order to determine the sequence of a particular segment of DNA, it is necessary to isolate that sequence in a pure form and in a reasonable quantity (micrograms). The enormous complexity of human and other eukaryotic genomes makes direct isolation of a particular sequence of DNA from cell nuclei technically impossible. (For example, if total human DNA consisting of 3.5×10^9 base pairs is digested with a restriction endonuclease and the particular DNA segment of interest is 1000 base pairs long, there may be a million DNA fragments of this size in the digest, all with different sequences.) Since the mid 1970s, it has been possible to manipulate DNA and to produce large quantities of pure DNA from complex genomes. The procedures which are used are collectively known as 'recombinant DNA' technology or 'genetic engineering'. These techniques enable pure genes to be isolated, transferred from one organism to another, and even 'tailored' to make possible the production of medically useful biological substances.

In order to produce workable quantities of a specific DNA fragment, the fragment is introduced into a bacterial cell where it is replicated many times, i.e. it is 'cloned'. The bacterial colony containing the introduced DNA fragment is a 'clone' and a collection (usually many thousands) of clones each containing different specific DNA fragments is termed a 'library'. The fragment is introduced into the host bacterial cell as part of the DNA of a 'vector'. Vectors are usually either 'plasmids' or bacterial viruses (bacteriophages). A plasmid is a naturally occurring closed circular DNA molecule which is capable of replicating independently within a bacterial host. Plasmids are between 1 and 30 kbp in size and generally encode proteins which are not essential for bacterial growth.

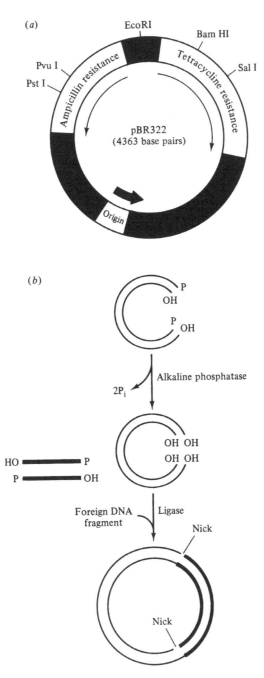

Fig. 7.9. (*a*) Genetic map of pBR322. Note that insertion of foreign DNA into the EcoRI site will not alter either the ampicillin or tetracycline resistance genes. Insertion at the Sal I or Bam HI sites

These proteins often determine resistance to antibiotics, and antibiotic resistance can be transferred between bacterial strains by plasmids. A bacteriophage which is commonly used in gene cloning is bacteriophage λ (lambda). Both with plasmids and with bacteriophages, DNA fragments are incorporated into the DNA complement of the vector before being transfected into bacterial cells. These are therefore 'recombinant' DNA molecules. Many different types of plasmids and bacteriophages have been produced for use in gene cloning.

There are two main types of library, a 'genomic' library containing genomic clones and a 'cDNA' library containing cDNA clones. Genomic clones contain fragments of genomic DNA, i.e. whole cellular DNA is cut into suitably sized fragments and each fragment is cloned. These DNA fragments therefore contain intron sequences, repeated sequences, coding sequences, etc., and are representative of all the DNA in the cell. cDNA clones, however, contain DNA fragments which have been copied from cellular mRNAs and contain protein-coding sequences (and 5' and 3' untranslated regions) from which intron sequences have been excised. Genomic and cDNA libraries can be prepared using either plasmids or bacteriophages as vectors.

A very commonly used plasmid is pBR322 (or derivatives of it). This is 4363 bp in length and contains two genes which confer ampicillin resistance and tetracycline resistance (Fig. 7.9(*a*)). Within these genes are unique sites for restriction endonucleases (Pst I, Bam HI, EcoRI, Hind III, etc.). The double-stranded DNA to be cloned (whether a genomic fragment or a cDNA) is cut with a suitable restriction endonuclease (e.g. EcoRI) and is mixed with pure plasmid DNA which has also been cut with EcoRI (Fig. 7.9(*b*)). Since the ends of the DNA are compatible, the cut DNA fragment anneals to the cut ends of the plasmid DNA and can then be covalently joined into the circular plasmid DNA by

Fig. 7.9. (*cont.*)
leads to inactivation of the tetracycline resistance gene, and at the Pst I or Pvu I sites inactivation of the ampicillin resistance gene. Only a few of the unique restriction sites are shown. The thin arrows show the direction of transcription of the antibiotic resistance genes. The thick arrow shows the direction of DNA replication. (*b*) After restriction enzyme cleavage of the plasmid, it is treated with alkaline phosphatase to prevent recircularization unless the foreign DNA fragment is inserted. After DNA ligase treatment, the remaining nicks in the recombinant plasmid are repaired by the host bacteria following transformation. Diagram modified from Old, R. W. & Primrose, S. B. (1985). *Principles of Gene Manipulation* (3rd edn). Blackwell Scientific Publications: Oxford.

DNA ligase, an enzyme which forms a covalent phosphodiester band between adjacent free 3′-OH and 5′-phosphate groups in DNA. The plasmid DNA is then introduced into a suitable strain of *E. coli*. Because the foreign DNA segment is inserted into the tetracycline resistance gene, bacteria containing such plasmids are now sensitive to tetracycline but still resistant to ampicillin since the ampicillin resistance gene is unaltered. When the transfected bacteria are grown on agar, bacterial colonies which are sensitive to both antibiotics contain no plasmids, those which are resistant to both contain unaltered plasmids, but those which are resistant to ampicillin but sensitive to tetracycline contain recombinant plasmids. These can be selected from the library and grown up.

The original foreign DNA being cloned was a mixture from many different DNA fragments, and therefore a method is needed to select which clones contain the sequence of interest. This is most commonly done by '*in situ* hybridization', which involves the application of a radioactive 'probe' (a nucleotide sequence complementary to the sequence of interest) to the collection of bacterial clones. Only clones containing the right sequences will bind the probe and these will show up as black spots after autoradiography (Fig. 7.10). In practice, this process (which is known as 'screening' the library) involves transferring plasmid or bacteriophage DNA to a nitrocellulose membrane before applying the probe; DNA sticks avidly to nitrocellulose but is still capable of annealing with complementary sequences. Cloning in pBR322 or its derivatives is widely used (apart from constructing libraries) to grow up large quantities of pure DNA fragments which have been initially cloned in other vectors. On the whole, pieces of DNA greater than 10 kb do not clone well in plasmids.

Figure 7.11 shows the life cycle of bacteriophage λ. This bacteriophage can accept pieces of DNA up to about 7.6 kbp and cloning in bacteriophage λ has the advantage of a much higher efficiency of transfection of the bacterial host compared to plasmids. λgt10 is now commonly used for the construction of cDNA libraries, this is shown schematically in Fig. 7.12.

For some purposes, vectors are required in which the foreign DNA insert can be transcribed and translated into protein. Such vectors are called expression vectors and are necessary if recombinant DNA techniques are being used to make a large amount of a mammalian protein in bacteria for subsequent therapeutic purposes and also if a DNA insert is being selected by its ability to produce a specific protein. In bacteria, translation commences before transcription is complete and there is no reason why foreign DNA cannot be read to give a protein product so long

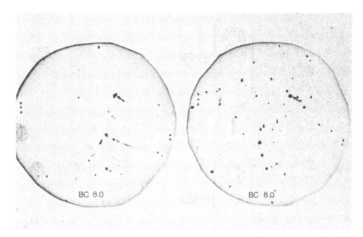

Fig. 7.10. Screening a library with an oligonucleotide probe. Many thousands of bacterial colonies are plated out on an agar culture medium and overlaid with a nitrocellulose membrane. DNA from each colony sticks to the membrane. After suitable treatment, the membrane is hybridized to a radioactive oligonucleotide probe whose sequence is complementary to the DNA sequence being sought. (The oligonucleotide probe is produced synthetically and its sequence is deduced using the genetic code from a knowledge of the amino acid sequence of the protein being investigated. A sequence of 5–10 amino acids is enough to produce a specific probe.) Unreacted probe is washed away and the filters are exposed to X-ray film. The procedure is usually carried out in duplicate to cut down the possibility of 'false positives'.

as the correct genetic elements are present. This is not a trivial problem and requires that the foreign DNA is inserted into the correct reading frame between a suitable promoter sequence and sequences coding to stop translation and terminate transcription. One experimental approach has been to produce vectors that allow insertion of the foreign DNA fragment at the end of a gene already coding for a protein. If the fragment is expressed, the transformed bacterial cell will produce a fusion protein containing a terminal amino acid sequence complementary to the DNA insert (Fig. 7.13). Such fusion proteins often precipitate within the cell, though this can be useful since it apparently renders them less likely to be proteolysed. A more promising approach for making large amounts of foreign protein by bacteria is to use expression plasmids which can grow in a bacterial strain that is capable of protein secretion, since in this way the protein may be protected from intracellular damage and easily

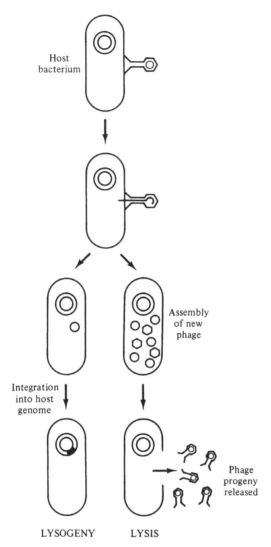

Fig. 7.11. The growth of phage λ. When it enters the host bacterium, the phage DNA can either replicate autonomously and lyse the host (lysis) or insert into the host chromosome and be propagated with it (lysogeny). The lysogeny decision can be reversed by, for example, irradiation of the bacterium with ultraviolet light. The alternative modes of growth are controlled by a switch mechanism operated through the action of regulatory proteins.

purified. Much work is currently being conducted on secretion vectors for use in *Bacillus subtilis*.

Both plasmids and bacteriophages have been used as expression vectors, an example of a high efficiency plasmid vector (pEX) is shown in Fig. 7.13(*a*). A commonly used bacteriophage expression vector is λgt11. Expression vector libraries are screened with antibodies against the protein of interest to identify positive clones (Fig. 7.13(*b*)).

Although bacterial expression vectors, whether plasmids or bacteriophages, lead to the synthesis of the required protein sequence, the machinery for correct post-translational processing of the polypeptide (especially glycosylation) is absent in the bacterial host. This lack of post-translational processing may make the protein biologically ineffective (particularly if it is being produced for therapeutic purposes).

Expression vectors based on viruses infecting animal cells (which can correctly process the synthesized protein) are now available, and these are being used commercially to produce clinically valuable therapeutic agents.

DNA sequencing

Until recently, the problem of determining the order of the individual bases in DNA and RNA presented a formidable chemical problem. Initially, methods were developed which were conceptually similar to methods of protein sequencing, i.e. the nucleic acid was degraded (usually by enzymes) into small fragments which were separated and the sequence of each fragment determined chemically. Deduction of the sequence of the parent molecule depended on obtaining the sequence of overlapping fragments derived from it. Although some small RNA molecules (e.g. tRNAs 75 nucleotides long) were completely sequenced by such methods, the smallest DNA molecule (that of the bacteriophage φX174) is about 5000 base pairs and new methods were needed to sequence such long stretches of nucleic acid. There are now two main methods of DNA sequencing, the Sanger didesoxy chain termination method and the Maxam–Gilbert chemical cleavage method. Both of these methods depend upon the ability to separate single strands of DNA differing by only one nucleotide in length. This is done by electrophoresis in polyacrylamide gels in a similar fashion to the separation of proteins (see p. 19), except that the DNA strands are separated in thinner gels using high currents and in the presence of urea, i.e. under 'denaturing' conditions. Such gels can separate DNA molecules from 2 to about 5000 nucleotides in length, each molecule differing by only one nucleotide (see Fig. 7.14).

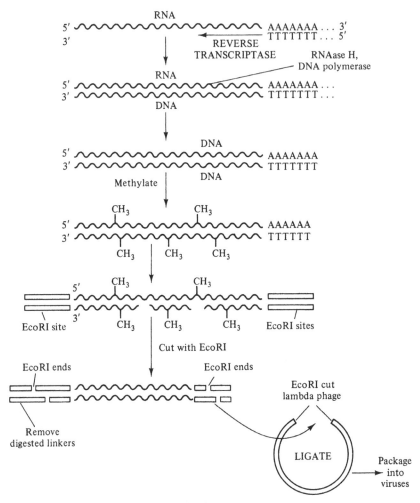

Fig. 7.12. Constructing a cDNA library in bacteriophage λ.
Total RNA is extracted from tissues or cells of interest and the small
proportion (1–3%) representing messenger RNA (mRNA) is
selected out by affinity chromatography on oligodT. (OligodT is a
synthetic oligonucleotide of repeating desoxythymidine residues –
i.e. dT–dT–dT–dT, etc. – which hybridize to the poly-A tails at 3'
ends of eukaryotic messenger RNAs. Messenger RNA molecules
can thus be purified specifically from a mixture of total RNA
molecules.) This, with an oligodT primer, is then copied into a
DNA strand by reverse transcriptase. The resulting RNA:DNA
hybrid duplex is digested with RNAase H (which preferentially
digests RNA in mixed RNA:DNA hybrid duplexes) and
simultaneously the RNA strand is replaced by a complementary
DNA strand by DNA polymerase. The ends of the DNA:DNA

The Sanger didesoxy method is a 'copying' procedure. A 'primer', i.e. a polynucleotide about 15–20 bases long, is annealed to a single strand of DNA (Fig. 7.14(*a*)). The enzyme DNA polymerase (which normally replicates DNA *in vivo*) will add nucleotides to the free 5′ end of the primer and will produce a new, exactly complementary DNA strand. To do this, the enzyme requires all four desoxyribonucleotide triphosphates (dATP, dTTP, dCTP and dGTP). If, however, didesoxyribonucleotide triphosphates (ddATP, ddTTP, ddCTP, ddGTP) are supplied, the enzyme will also incorporate these. Since didesoxynucleotides lack a 3′-OH group, no further nucleotides can be added and chain termination occurs. DNA polymerase is therefore supplied with a mixture of all four desoxribonucleotides and additionally one of these in the didesoxy form. If, for example, the mixture is dATP, dCTP, dGTP, dTTP with ddCTP added, DNA polymerase will copy the template strand until a C needs to be incorporated and will then incorporate either a dCTP (in which case further copying occurs until the next C is needed) or a ddCTP (in which case chain termination occurs and new strand synthesis stops). The end result is a mixture of new DNA strands of different lengths, all ending in a 3′-incorporated ddCTP. The overall lengths of the strands depend on the relative concentrations of dCTP and ddCTP supplied: if the ddCTP concentration is high, chain termination occurs as soon as a C is reached; if it is low, chain termination is rare and the new strands are much longer. By judicious adjustment of the concentrations of dCTP and ddCTP, chain termination can occur at all possible C positions and thus the whole range of possible new strand lengths results. By carrying out four parallel incubations, each with four desoxynucleotides and a different didesoxy-

Fig. 7.12. (*cont*)

helices are now 'filled in' to produce 'blunt' ends. The DNA is then treated with S-adenosylmethionine and EcoRI methylase, which methylates any cytosines in EcoRI recognition sequences within the DNA and therefore protects the DNA from digestion with EcoRI. EcoRI linkers, i.e. synthetic stretches of DNA (~10 bp long) containing the EcoRI recognition sequence, are then ligated to the end of the DNA and the ligated molecules digested with EcoRI, leaving the DNA with EcoRI sites at the 5′ and 3′ ends. After removal of undigested linkers, the DNA is ligated into λ*gt10* previously cut with EcoRI. This λ*gt10* DNA is then 'packaged' into active viruses and used to infect the appropriate host bacterium. The resulting library can contain several million independent recombinants (i.e. bacteria infected with individually different bacteriophages containing a different DNA insert) per microgram of initial mRNA used.

(*a*)

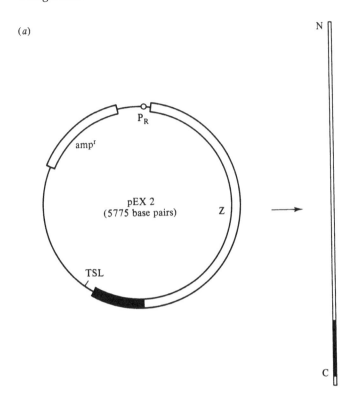

(*b*)

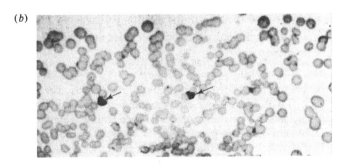

Fig. 7.13. (*a*) The bacterial expression vector pEX. The pEX vector expresses a β galactosidase hybrid protein which is insoluble, forming dense inclusion bodies in the cell accounting for up to 25 % of total *E. coli* protein. The plasmid contains an ampicillin resistance gene (amp^r), a strong promotor (P_R), a modified β galactosidase gene sequence (Z), an oligonucleotide linker containing cloning sites, L, which in the diagram contains a short open reading frame cDNA (solid), an oligonucleotide containing translation stop codons in all reading frames, S, and transcription terminator sequences, T,

nucleotide (ddATP, ddCTP, ddGTP or ddTTP), four different mixtures of new strands are produced, each mixture containing strands terminating in either didesoxy A, C, G or T. These mixtures can then be separated in four parallel lanes on a polyacrylamide gel. Since the smallest strands are at the bottom and largest at the top and each strand differs by one nucleotide, the sequence can be read by beginning at the bottom and then reading towards the top switching from lane to lane according to where the next largest fragment is found (Fig. 7.14). In order to visualize DNA strands in the gel one of the desoxyribonucleotides is supplied radioactively labelled in its α-phosphate. After electrophoresis, the gel is exposed to X-ray film to produce the result in Fig. 7.14). In practice, the DNA polymerase used is 'Klenow fragment' which is a fragment of *E. coli* DNA polymerase lacking the 5'- to 3'-exonuclease activity found in the intact enzyme. Frequently, instead of an α-^{32}P isotope, desoxyribonucleotide, an α-^{35}S derivative is used. This is incorporated almost as readily as the α-^{32}P form, but has the advantage that the ^{35}S isotope has a much longer half-life than ^{32}P and also is a weaker β-particle emitter. This means that, during autoradiography, the local dose of radioactivity in the gel is lower and therefore closely spaced bands (especially in the upper regions of the gel) can be resolved more easily.

The didesoxy chain termination method prefers a single-stranded DNA template. To produce this, use is made of a single-stranded filamentous phage M13 (Fig. 7.15). This infects a suitable *E.coli* host and its single-stranded DNA genome is converted into a double-stranded replicative or 'RF' form. One strand of this is copied many times to produce several hundred progeny single-stranded phages which are

Fig. 7.13. (*cont*)

from phage fd. The expressed polypeptide corresponding to the hybrid gene and cDNA insert is shown on the right. The plasmid has been constructed in three versions each differing by one base pair to maximize the chances of inserting a foreign cDNA fragment in the correct reading frame. Grown in the correct host, expression of the fusion protein can be temperature-controlled. Thus, the bacteria multiply and the plasmid is amplified at 30 °C. On altering the temperature to 42 °C, the strong promoter ensures that a massive amount of fusion protein is rapidly made.

(*b*) The expression vector library is screened with antibodies to detect positive clones. Shown is a human liver mRNA library screened with an antibody to human serum albumin. Positive clones (arrows) are detected by using anti-immunoglobulin antibodies coupled to horseradish peroxidase (cf. Western blotting p. 19, and immunohistochemistry p. 55).

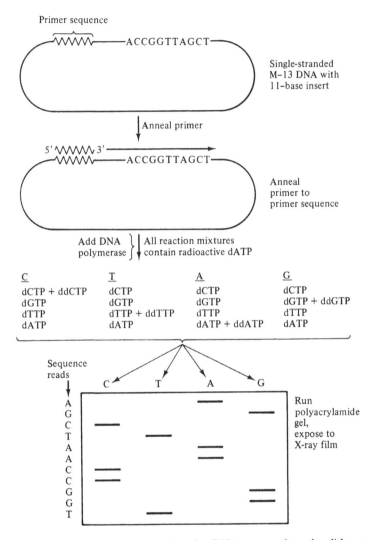

Fig. 7.14. The Sanger procedure for DNA sequencing: the didesoxy method. Chain termination is enforced by including didesoxy nucleotides in the reaction mix. Newly synthesized strands, differing by as little as one nucleotide in length, can be separated in polyacrylamide gels. The sequence is then read off as shown. See text for further details.

extruded from the host bacterial cell without lysis. These single-stranded bacteriophages can be harvested from the supernatants of bacterial cultures and the protein stripped from them by extraction with phenol to produce single-stranded circular DNA which is suitable for didesoxy

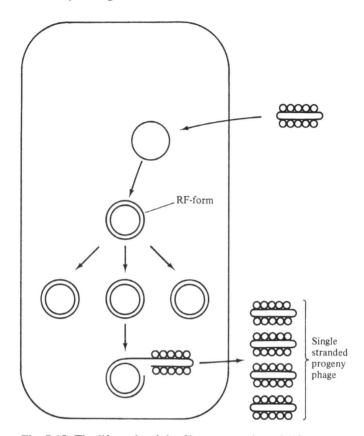

Fig. 7.15. The life cycle of the filamentous phage M13:
1   The single-stranded form of the phage is infective and enters the
bacterial host cell via the pilus.
2   Single-stranded DNA is converted to the double-stranded,
replicative or RF form.
3   Replication of the RF form gives over 100 progeny per cell.
4   Single strands of DNA are produced by the rolling ball method.
5   Over 200 single-stranded progeny per cell generation are
produced and released into the medium.

Foreign DNA inserts can be ligated into the RF form (restriction
endonucleases work only on double-stranded DNA), and, after
transformation, many single-stranded phages containing copies of
the inserted foreign DNA result. This is a simple way of producing
large quantities of the ideal template for the didesoxy sequencing
method.

sequencing. The M13 bacteriophage has been genetically engineered so
that a piece of DNA has been inserted into a region unimportant for viral
replication. This piece of DNA contains several unique restriction sites
which can be used to insert double-stranded DNA cut out of other vectors

using similar restriction endonucleases. For example, a DNA fragment excised with EcoRI and Pst I can be inserted into the RF from M13 cut with EcoRI and Pst I. This can then be used to infect *E.coli* and to produce many single-stranded DNA molecules containing one strand of the insert as part of their sequence. M13 also has a special sequence (ACTGGCCGTCGTTTTAC) inserted 3′ to the restriction endonuclease sites. This means that a 'universal' primer, i.e. a synthetic 17-base polynucleotide complementary to this sequence, can be annealed to the single-stranded DNA. Polymerization by Klenow fragment polymerase then proceeds from the end of this primer into the region containing the inserted 'unknown' DNA. M13 has been engineered so that different strains of the bacteriophage RF forms have the added restriction endonuclease sites in opposite orientations, i.e. a piece of inserted DNA is ligated into one M13 strain in one direction and into another in an opposite direction. By using both strains for sequencing, the sequence of the unknown DNA is therefore read on both DNA strands. This is an important check in making sure the derived sequence of bases is correct. While it is possible to put large fragments (of more than 5 kb) of DNA into M13, fragments of 300–900 bases are usually inserted. These tend to be more stable and to fit in with the resolving power of the usual sequencing gels (up to about 500 bases).

The Maxam–Gilbert method similarly requires single-stranded DNA (but produced by different means) and also separates radioactive DNA fragments on polyacrylamide gels. Instead of chain termination reactions, however, different lengths of DNA are produced by specific chemical reactions which cleave at A, G, C+T or C residues.

These sequencing techniques make it feasible for one worker to produce up to 1000 bases of sequence in a day. The generation of such rapid sequence requires storage and manipulation by computers. These are used to record and retrieve sequences, to check data bases for similar or identical sequences, to order large stretches of sequences over lengths of several hundred kilobases, to search for particular restriction endonuclease sites, and to 'translate' automatically the DNA sequence into the respective amino acid sequence, i.e. to look for 'open' reading frames. Newer automated DNA sequencing machines based on the above sequencing technologies are now coming into existence which will make possible much higher rates of DNA sequence accumulation (possibly up to one million bases per day). It is likely that the whole human genome – all 3.5×10^9 base pairs – will be sequenced within the next decade.

Sequence organization of the human genome

DNA sequences in the human genome can be grouped into three categories:

1. 'structural' genes,
2. repetitious DNA, and
3. unclassified or 'spacer' DNA.

'Structural' genes

These genes code for a recognizable protein (or a cellular RNA such as ribosomal RNA) and are present either as one or at most a few copies per haploid genome. An exception are the genes coding for tRNAs and ribosomal RNAs, which may be tandemly repeated many times per haploid genome. In addition, some 'single-copy' genes have closely related copies (pseudogenes) which may be close by or great distances away in the genome. 'Transcription units' coding for functional proteins probably make up less than 10 % of the human genome. Furthermore, only a minor proportion of each transcriptional unit consists of protein coding sequences.

Repetitious DNA

These DNA sequences are repeated many times (10^3–10^6 copies) per haploid genome and do not code for any known protein. Such repeated sequences may make up as much as 30–40 % of total human DNA. Although previous work studying the re-annealing of sheared denatured DNA provided clear evidence for the presence of several different classes of repetitive sequences in eukaryotic DNA, only with the advent of cloning and sequencing technology has it been possible to study the sequence organization of the human genome in detail. DNA sequences in the human genome can be classified as 'unique' (present in one or a very few copies in each genome), highly repetitive (present in approximately 10^6 copies per genome), and moderately repetitive (present in 10^3–10^5 copies per genome).

Highly repetitive sequences are characteristically short (5–10 bp) and repeated in tandem many thousands of times, i.e. they are not interspersed with different non-repetitive sequences. The sequence of each repeating unit is highly conserved. Since this can lead to a base composition significantly different from the bulk of the genomic DNA, this class of highly repetitious DNA can band separately on caesium chloride gradients and is then termed 'satellite' DNA. However, because some forms of this class of repeated sequence DNA can sediment with the bulk of the genome DNA and not form satellite bands, the term 'simple sequence repetitious DNA' is preferable. The human genome contains at least 10 different sequence families of this class of DNA, most of which are 5–10 bp, but human and primate DNA is unusual in also having a longer repeat sequence (>170 bp, the 'alpha sequence') among this class of repeated sequences. *In situ* hybridization experiments have shown that most (but not all) repeated sequences in this class are located in the centromere or telomere (chromosome tips) regions of metaphase chromosomes. For this reason, and also because simple sequence repetitious DNA does not code for any recognizable protein and does not appear to be transcribed into RNA, the function of this class of DNA is thought to be 'structural' and possibly to provide binding sites for proteins involved in 'organizing' chromosomes. If so, it is not clear why such large stretches of repeated DNA (amounting overall to about 10 % of the genome) should be required at centromere or telomere regions.

Moderately (intermediate) repetitive sequences can be classified as short (150–300 bp), amounting to up to 40 % of the genome, or long (around 5 kbp), amounting to about 1–2 % of the genome. In contrast to simple repetitive DNA, intermediate short (and long) repeat sequences are dispersed throughout the genome and not clustered into tandem repeats. One of the major families in this group is the Alu family, named because many (but not all) of its members contain recognition sequences for the restriction enzyme Alu I. By annealing experiments using DNA sheared into pieces about 5–10 kbp long and by direct visualization of double-stranded or single-stranded DNA molecules in the electron microscope (see Fig. 7.16(a)), it can be estimated that Alu sequences (totalling 5×10^5 members) occur on average every 5 kbp in human genomic DNA and make up 5–10 % of total DNA. Sequencing cloned genomic DNA containing Alu sequences has shown (again in contrast to simple repetitive DNA) that each member of the Alu family has a different but closely related sequence. This is 300 bp long with the 5′ end of the sequence (130 bp) showing close homology with the 3′ end, the latter having an insert of 30 bp (Fig. 7.16(b)). Alu sequences are found throughout the genome and can occur in introns or in flanking sequences of structural genes. Alu sequences can be found in hnRNA within introns and are normally spliced out. About 5 % of mRNAs, however, contain Alu sequences, and these occur in the 5′ and 3′ untranslated region of the message. A suggested function of Alu sequences is that they may represent initiation sites for DNA replication. This is based on the number of Alu sites present (which matches estimates of the number of replication origins on eukaryotic chromosomes) and the occurrence

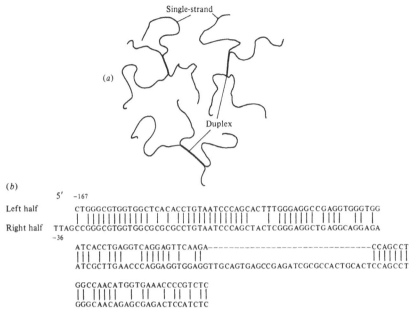

Fig. 7.16. Repeated sequences in human genomic DNA. (*a*) If DNA is sheared into large pieces and then denatured by heating, strands will reassociate into double helices on cooling provided opposite strands have complementary sequences. If reassociation conditions are chosen so that only sequences repeated many times in genomic DNA will reanneal, the DNA can then be examined under the electron microscope and an estimate made of how far apart along the length of the DNA the repeated sequences are. Under the electron microscope reannealed (double helical) DNA appears thicker than the intervening single-stranded DNA. Note pairing of repeated sequences can occur between different strands or within the same strand of DNA. (*b*) Human Alu sequences – the left half of the sequence is homologous to the right half but contains an additional 30 bp insert. Here the right half of the sequence is shown written 'in phase' under the left half of the sequence; the right half contains a 30 bp insert missing in the left half (broken line). Different human Alu sequences show overall about an 85% homology, similar families of sequences are present in other animal species.

within the consensus Alu sequence of a region similar to the sequence of the DNA replication origin seen in a number of DNA viruses.

Human DNA (along with other mammals) also contains families of longer (5–7 kbp) dispersed repeats, which make up about 1–2 % of the genome. One prominent member of this family is the Kpn I dispersed repeat sequence which can be released as 5–7 kbp fragments by the restriction enzyme Kpn I. In some cases (notably yeast and *Drosophila*),

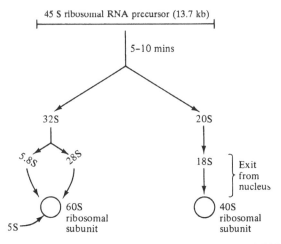

Fig. 7.17. The maturation pathway of ribosomal RNA. The human rRNA gene is 13.7 kb in length and codes for 18S, 5.8S and 28S rRNA. The fourth ribosomal RNA (5S) is synthesized from an entirely separate transcription unit by RNA polymerase III. RNA polymerase I transcribes the rRNA gene into a 45S primary transcript which is then processed in the nucleolus to produce mature ribosomal RNAs. The whole maturation process takes about 30 min.

the sequence of these long, dispersed repeats shows homology with retroviruses, in particular showing flanking 300–600 bp direct repeats similar to the long terminal repeats (LTRs) seen in integrated retroviruses.

Unclassified or 'spacer' DNA

The overall picture emerging of the human genome is one in which functional transcriptional units producing proteins are separated by great distances of DNA (e.g. 20–100 kbp). The intervening DNA does not code for any known protein or RNA, may or may not contain elements of repeated sequences, and appears functionless.

Transcription and transcriptional units

Transcription (copying) of DNA into RNA is carried out by RNA polymerases which are complex multi-subunited enzymes and are of three types in eukaryotic nuclei (I, II and III).

RNA polymerase I is located in the nucleolus and transcribes ribosomal RNA genes. In Man, there are approximately 200–300 copies of the ribosomal RNA transcription unit within the nucleolus, each with an identical sequence coding for the 45S rRNA precursor. Separating each transcription unit are non-transcribed spacer regions varying from

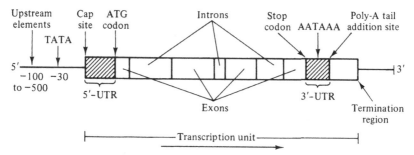

Fig. 7.18. The structure of a typical RNA polymerase II transcription unit. See text for details.

20–30 kb in length. The 45S rRNA precursor molecule is processed into 28S, 18S and 5.8S RNAs seen in the cytoplasmic ribosome. (Fig. 7.17).

RNA polymerase II is a nucleoplasmic enzyme which copies transcriptional units containing protein-coding information. The structure of a 'typical' RNA polymerase transcription unit is shown in Fig. 7.18. The protein-coding sequence begins at the 5' end with the start codon AUG (= methionine) and ends at a stop codon (UAA, UGA or UAG) (for convenience the RNA form of the genetic code is adhered to). At intervals, the protein-coding sequence is interrupted by 'intervening sequences' or 'introns'. These are present in nearly all the eukaryotic structural genes so far examined (a notable exception being those coding for histones from which they are absent) and their discovery in the late 1970s was a major surprise in molecular biology. Introns, which, although transcribed, are removed from the RNA transcript and destroyed in the nucleus (as opposed to exons which leave the nuclei as parts of tailored protein-coding transcripts), can make up as much as 80–90 % of the total transcriptional unit in some genes. The intron/exon sequence boundary is conserved (presumably as an essential part of the splicing mechanism). There are two possibilities to explain the evolution of introns: the first is that primordial structural genes lacked introns and therefore resemble bacterial structural genes; the second is that the first genes contained segmented coding sequences interrupted by introns and that present-day intronless genes in lower organisms represent an evolutionary 'advance', in that intron loss could lead to a streamlining of the machinery for the replication and transcription of DNA. Close examination of the intron positions of the genes coding for proteins which have been highly conserved throughout evolution (e.g. enzymes of glycolysis) currently favours the second possibility.

In addition to the DNA sequence between the translation initiation and termination codons (i.e. the sequence containing exons and introns), additional upstream and downstream sequences are present which can appear in the final mRNA product and must be considered part of the functional 'gene'. Up to approximately 50 nucleotides upstream (5') from the start of the protein-coding sequence is the site at which RNA polymerase initiates transcription. This is known as the 'cap' site because the initiating nucleotide (usually A or G) is 'capped' with 7-methylguanosine (Fig. 7.19). This occurs very rapidly during transcription, when the RNA transcript is about 20–30 nucleotides long. Approximately 30 nucleotides further upstream from the initiating cap site occurs a highly conserved sequence of TATA (the TATA box) which is essential for the correct positioning of RNA polymerase II to begin transcription at the cap site. Further upstream still (at about −110 and −40) occur upstream elements with less highly conserved sequences which also appear to be important in producing high rates of transcription. These upstream elements and the TATA box are thought to act in the correct binding of RNA polymerase II to begin transcription at the cap site. In addition to these sequences, remote DNA sequences (which can be several thousand bases distant from the initiation site and in either orientation to the transcription unit, i.e. 5'–3' or 3'–5') are known to exist which can have dramatic effects in increasing transcription rates of particular genes.

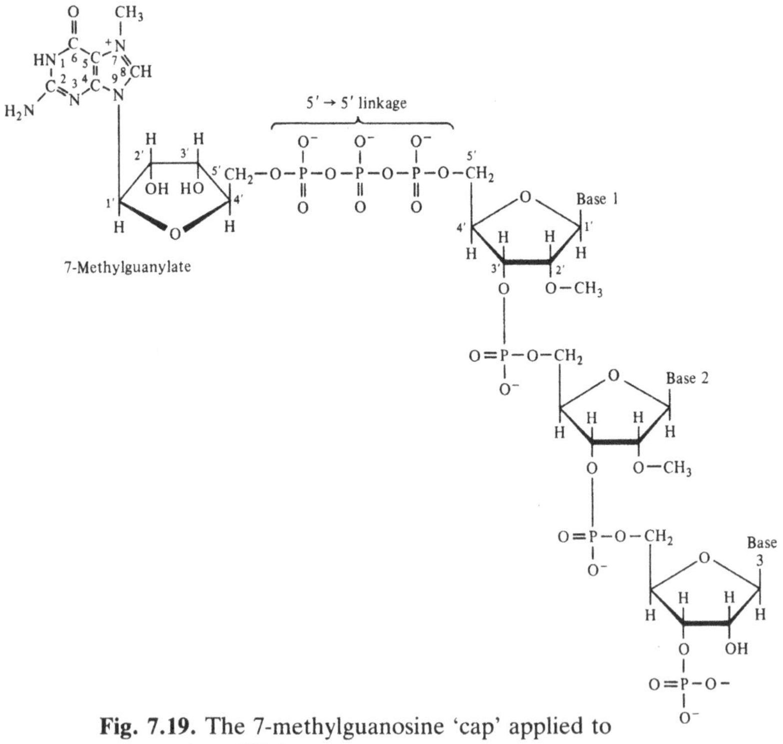

Fig. 7.19. The 7-methylguanosine 'cap' applied to eukaryotic mRNAs.

These DNA elements are termed 'enhancers' and the way in which they act 'at a distance' is not currently clear, although it is possible that their effect is produced by long-distance alterations of chromatin structure allowing access of transcription factors to active genes.

Downstream (3') from the codon which signals the end of translation into protein (UAA, UAG or UGA) occurs a stretch of DNA which is transcribed into the mature mRNA but not translated (the 3'-untranslated region). This can be 1000–2000 nucleotides in length. RNA polymerase II continues past the protein-coding sequence for a variable distance until it transcribes through a sequence AATAAA which is highly conserved in eukaryotic mRNAs. This sequence is the signal for the addition of a poly A 'tail' of about 250 nucleotides, the addition occurring about 20 nucleotides further downstream from the AATAAA signal. RNA polymerase II continues transcribing past the polyadenylation signal and the polyadenylation site (20 nucleotides further on) for up to 2 kb before terminating transcription (the exact sites and signals for this are currently unknown). 'Excess' RNA transcribed past the polyadenylation site is removed and destroyed and the poly A tail is rapidly (within seconds) added to the 3' end of the transcript. Termination of transcription is therefore beyond and distinct from the polyadenylation site. The immediate primary transcript (hnRNA) from an RNA polymerase II transcription unit therefore has a capped 5' and a variable (e.g. 5–50 nucleotide) untranslated 5' region, a protein coding region consisting of exons separated by a variable number of introns, a 3'-untranslated region (which can be a few thousand nucleotides long) and a 3' poly A tail of about 250 nucleotides.

RNA polymerase III is a nucleoplasmic enzyme responsible for the transcription of genes coding for tRNAs and other small RNAs (<300 nucleotides long) in eukaryotic cells. tRNAs average 75–80 nucleotides in length and are coded for by approximately 1300 tRNA genes in the human genome. (Remember that several tRNAs may be used for the same amino acid and the total number of copies of genes coding for tRNAs may represent 40 or 50 individually different molecules.)

Complex transcriptional units

The transcription unit described above can be classified as 'simple' because it leads to the production of only one species of mRNA transcript which, on translation, produces a single species of protein. The β-globin gene is an example of a simple transcription unit. Many instances are now known, however, of transcription units which are 'complex', i.e. they lead to the production of more than one type of mRNA transcript and to the production of more than one type of protein. Complex transcription

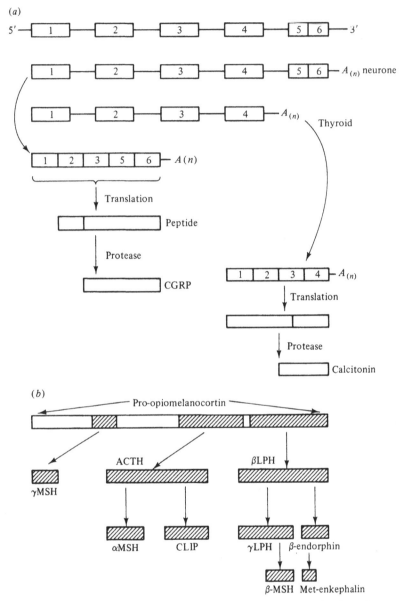

Fig. 7.20. Complex transcriptional units and the production of alternative protein products from the same gene. (*a*) The mechanism of production of calcitonin and CGRP. The thyroid uses the first polyadenylation site (A_n), the brain uses the other, and calcitonin and CGRP are produced. (*b*) An extreme example of one primary protein sequence giving rise to several different polypeptide hormones, such as adrenocorticotrophin (ACTH), lipotrophic hormones (LPH), melanocyte stimulating hormone (MSH), etc.

units initiate transcription at one site but either have two or more alternative poly A addition sites or have a single poly A site but employ differential splicing of the initial primary transcript. Alternative poly A sites may be combined with differential splicing. An example of this process is shown in Fig. 7.20, which illustrates the differential processing of the calcitonin gene to produce either calcitonin, a hormone which lowers the concentration of calcium in the blood, or calcitonin-gene-related-peptide (CGRP), which is possibly involved in taste.

A different method of producing alternative protein products from the same gene is to process the polypeptide product rather than the RNA transcript. An extreme example of this is the polypeptide pro-opiomelanocortin which is produced by the anterior pituitary. Proteolytic cleavage of this leads to the production of nine distinct peptides, most of which are known to be biologically active (Fig. 7.20).

Control of gene expression
Each cell in the human body contains its own complement of proteins, which are estimated to be at least 5000 and possibly 30 000 in number. Some of these proteins are present in much greater concentrations than others, and a proportion of these (perhaps a few hundred) are cell-specific, i.e. they enable the cell to function according to its particular cell type and role in the body. A cell also contains many different individual mRNAs present in widely varying numbers of copies. It has been estimated that a typical liver cell contains about 10^6 mRNA molecules, the most abundant 100 molecules of which are present in about 5000–50 000 copies per cell, and the least abundant 10 000 molecules of which average between 0.1 and 10 copies per cell. Cells can rapidly 'adapt' from a steady-state condition in response to hormonal, nutritional or other signals, such as growth factors. For example, starvation leads to the induction (within hours or a few days) of the enzymes of gluconeogenesis (conversion of amino acids to glucose) in liver cells. The cytochrome P450 system, which is responsible for the detoxification of drugs such as barbiturates and morphine and also of various carcinogens, is also rapidly induced in liver cells on environmental exposure to these agents.

The possible control points at which genes and their products can be up- or down-regulated are at the level of transcription, during mRNA processing, during transport out of the nucleus, by affecting the stability of the mRNA in the cytoplasm, or by controlling translation of the message into a polypeptide chain. It is now widely believed that the majority of protein-coding genes are regulated at the transcriptional level, although specific examples of gene control at other sites (listed above) have been quite clearly demonstrated. Factors which act on genes to increase the rate of transcription either have to be very close to the gene to exert their effect (acting in '*cis*') or can act at a distance (acting in '*trans*'). It is assumed that regulatory proteins exist in eukaryotes (as they do in prokaryotes) which can bind to specific DNA regions or to RNA polymerase molecules and control transcription rates, and such regulatory proteins are being characterized. 'Active' genes differ from 'inactive' genes by being more sensitive to digestion with enzymes that digest DNA (particularly 'hypersensitive' sites being found at the 5' end of the active gene), by being 'undermethylated' (i.e. containing less methylated cytosine than the inactive version of the gene) and by being associated with increased amounts of non-histone proteins which are thought to be involved in gene regulation. It is also widely assumed that activation of specific genes requires 'loosening' of local chromatin structure to allow access of polymerases and other 'factors' (see Fig. 7.4).

Much less emphasis is currently placed on gene regulation by non-transcriptional mechanisms. While differential polyadenylation and differential splicing are known to occur (see above), the factors which control this are presently poorly understood. A clear example of cytoplasmic control is provided by prolactin, a hormone which increases casein (milk protein) production in breast cells. Prolactin increases the number of casein mRNAs in these cells by about 300-fold, largely by increasing the half-life of the mRNA in the cytoplasm by 30–50 times. Translational control, i.e. increasing the rate of translation of an individual mRNA directly (rather than increasing the number of copies or making the mRNA more stable so that it can be copied more often as with prolactin) has been clearly demonstrated in lower vertebrates, but its significance in human cells has yet to be established.

Splicing of primary RNA transcripts

All primary RNA transcripts, with the exception of two types produced by RNA polymerase III (5S RNA and possibly small nuclear RNAs) undergo some form of processing (splicing) which removes internal portions of the sequence to produce the 'mature' RNA. The simplest example is the excision of the short (14-base) intervening sequence which interrupts the sequence one base 3' to the anticodon in some tRNAs in some organisms. There does not appear to be any consensus sequence in tRNA introns, and accurate positioning for the splicing enzymes appears to depend on the tertiary structure of the tRNA as a whole. Nuclear hnRNA introns have an invariant sequence:

$$5'-{}^{A}_{C}AG \quad \Big| \quad \overset{\leftarrow \qquad \text{intron} \qquad \rightarrow}{\underline{GT} \qquad\qquad (Py)bX\underline{CAG}} \quad \Big| \quad G{}^{G}_{T}-3'$$

such that GT occurs at the 5′ end of the intron at the exon/intron boundary and AG at the 3′ end of the intron. At the 5′ exon/intron boundary, cleavage leads to a 3′-hydroxyl on the 3′ end of the exon and a 5′-phosphate on the G at the 5′ end of the intron. The next step is to form a branched structure (as shown in Fig. 7.21) between this 5′-phosphate and the 2′-hydroxyl of an adenylate residue lying 10–30 residues from the 3′ intron/exon junction within this consensus sequence:

$$CU\left\{{}^{C}_{A}\right\} A \left\{{}^{C}_{U}\right\}$$

Finally, the 3′ intron/exon junction is cleaved past the AG sequence and the 5′ phosphate left on the nucleotide immediately 3′ to the G is joined to the free 3′-hydroxyl on the 3′ end of the adjacent exon. Splicing of

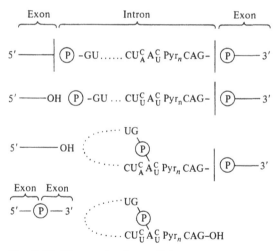

Fig. 7.21. The mechanism splicing remaining intron sequences from nuclear RNAs. See text for details. The linkage between the guanylate (G) at the cut 5′ end of the intron and the adenylate (A) close to the 3′ end of the intron is 5′–2′. The adenylate retains its usual 5′–3′ and 3′–5′ linkages (Fig. 7.1). The phosphate finally linking the two exons is derived from the first nucleotide of the 3′ exon.

hnRNA into the mature mRNA takes 5–20 minutes and is exclusively nuclear. Finally, the mature mRNA (with its 5' 'cap' and 3' poly A tail) passes into the cytoplasm (probably via the nuclear pores) for translation on ribosomes. The half-life of different mRNAs in mammalian cells shows considerable variation e.g. from a few hours or less to several days. The 3' poly A tail, by a mechanism which is not understood, appears to confer stability on the mRNA in the cytoplasm.

Systemic lupus erythematosus (SLE) is a human disease presenting with arthritis, skin rashes, and various kidney and connective tissue disorders. SLE belongs to a group of diseases known as 'autoimmune' disorders because antibodies arise (apparently spontaneously) against normal body constituents. It has been known for many years that patients with SLE have circulating antibodies against components of the cell nucleus. Some of these antibodies are directed against DNA, others against nuclear proteins, e.g. histones. Two common nuclear antigens against which SLE serum antibodies are directed are the RNP (ribonucleoprotein) antigen and the 'Sm' antigen. The RNP antigen is so-called because its immunoreactivity is destroyed by RNAase and by trypsin; the Sm antigen is named after the SLE serum with which it was originally detected. These antigens are conserved during evolution and SLE antibodies will recognize them in cell nuclei from lower species. Eukaryotic nuclei contain small nuclear RNAs (56–217 nucleotides in length), abbreviated as snRNAs. These are known to be contained in small nuclear ribonucleoprotein particles (snRNPs), each containing one RNA molecule and approximately 10 different proteins. There are at least six different types of snRNP, named U_1–U_6; some proteins occur in all snRNPs, others are unique to a particular snRNP. It is now known that the Sm antigen is a protein component of all six snRNPs, while the RNP antigen is contained in some snRNPs but not others. Why these components should become antigenic in patients with SLE is unknown. The role of snRNPs in the cell nucleus is now known to be in the splicing process described above, snRNPs forming part of 'spliceosomes'. These particles, which can be isolated from cell nuclei and visualized under the electron microscope, are about 25 nm across and 40–60 nm long. They contain variable numbers of snRNPs and hnRNA at various stages in the splicing process. The individual roles of each snRNP in splicing are being clarified, for example, U_1snRNP is involved in locating the 5' exon/intron boundary (Fig. 7.21) and the formation of the branched structure shown in Fig. 7.21 requires U_2snRNP.

Duplicated protein-coding genes and pseudogenes

Several proteins, especially abundant ones like actin and tubulin, occur in 'families' containing proteins with very similar but not identical amino acid sequences, and are often expressed in a tissue-specific manner. In vertebrates, actin genes are known to be duplicated 5–30 times, α- and β-tubulins 5–15 times, and histones over 100 times. An extreme example (necessary as part of the mechanism generating antibody diversity) is the duplication many hundreds of times of the genes coding for the variable portions of immunoglobulins. These genes have apparently arisen by duplication and then evolution into related but not identical protein-coding sequences. Furthermore, the organization of the transcriptional units, especially in terms of intron numbers and positions, is also usually preserved as evidence of the common origin of members of the duplicated gene family. These genes are functional and can be transcribed either in a tissue-specific manner or at different times during development (e.g. with the β-globin gene cluster).

A different type of gene 'family' is represented by pseudogenes. These were first recognized when the genes coding for 5S RNA in *Xenopus* were sequenced, when a gene was found which was considerably shorter and contained several base changes when compared to a normal functional 5S RNA gene, hence the term 'pseudogene'. This term is used to describe sequences which are clearly related to functional genes but which are 'defective' in that no recognizable functional protein or RNA product can be produced from the sequence (although this does not necessarily mean that the sequence is never transcribed). Pseudogenes can be classified into two main types, first those which retain the intron/exon arrangement of the functional 'parent' gene, and secondly, those which do not. The first category is best exemplified by the globin pseudogenes which are closely associated with functional globin genes but which, either because of the acquisition of stop codons or because of sequence changes which would prevent RNA processing, are now non-functional.

The second category are 'processed' pseudogenes, and show certain characteristics:

1. They lack intron sequences but now have nearly contiguous exon sequences.
2. They have poly A tails at the 3′ end.
3. They have direct repeat sequences which flank the 5′ coding region and the 3′ poly A tail, although these bear no resemblance to the sequences flanking similar regions in the 'parent' functional gene.

4. They are usually found on different chromosomes from the parent gene (i.e. they are not syntenic).

This class of pseudogenes appears to have been formed relatively recently in evolutionary time and integrated into the genome as a copy of the processed mRNA transcript from the functional gene sequence.

'Defects' in eukaryotic transcription units

The genes coding for the globin proteins of haemoglobin currently provide the best understood examples of the different ways in which alterations in a transcription unit can lead to defective protein products and hence to clinically significant disease states. The genes coding for α- and β-globin are localized on chromosomes 16 and 11, respectively, as large gene clusters which in the case of the β-globin genes, span up to 60 kbp of DNA. The order of genes in the α- and β-globin gene clusters is shown in Fig. 7.22. Note the presence of intron-containing pseudogenes in both gene clusters. During normal development, the genes in the α and β gene cluster are sequentially expressed to produce the tetrameric haemoglobin molecule typical of that period of maturation.

In the normal adult, haemoglobin is largely $\alpha_2\beta_2$ (HbA) with a minor component $\alpha_2\delta_2$ (HbA$_2$). Inherited defects in globin chain synthesis are either structural variants, i.e. single (largely) amino acid substitutions in the α- or β-globin chains, or imbalances of globin chain production (the

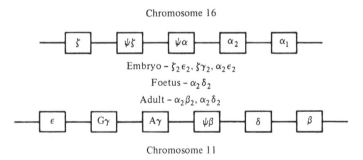

Fig. 7.22. The structure of the α- and β-globin gene clusters. Note the presence of two pseudogenes ($\psi\zeta$ and $\psi\alpha$) in the α-globin gene cluster and one ($\psi\beta$) in the β gene cluster. The other genes code for functional haemoglobin chains which are expressed at different times of development as shown. There are two genes coding for γ chains, one with glycine at amino acid position 136 (Gγ) and one with alanine (Aγ). There are two genes coding for α chains (α_1 and α_2). The major adult form of human haemoglobin is HbA ($\alpha_2\beta_2$) with a minor proportion of HbA$_2$ ($\alpha_2\delta_2$).

thalassaemias), or (more rarely and clinically less important) defects of the switch from foetal to adult globin chain synthesis.

At least 300 structural variants of haemoglobin are known, many of which are 'neutral' substitutions producing no clinical symptoms. The most important is HbS (sickle-cell haemoglobin) when valine has replaced glutamate at position 6 in the β-chain. This causes haemoglobin to polymerize at low oxygen tensions, leading to 'sickling' of red cells, anaemia (due to lysis and decreased survival of red cells), and to blockage of arteries which can deprive the bone marrow, brain, kidneys, etc., of their normal blood supply. HbC and HbE are also two common clinically important variants in Africa and South-east Asia, respectively. Point mutations leading to single amino acid substitutions in globin chains have in the past thrown important light on the genetic code and the primary and tertiary structure of proteins.

In recent years, the molecular defects underlying the α- and β-thalassaemias have been unravelled and have proved surprisingly complex compared to the clinical picture which can appear very similar from one case to another. The molecular defects which have been characterized in the α-thalassaemias are summarized in Fig. 7.23. In α^0-thalassaemias no α chains are produced, and the α-globin cluster shows deletion of various portions of DNA which can extend from the αgene through to the δ gene. In α^+-thalassaemias, some α chains are produced but at a greatly reduced rate. The molecular lesions here have been shown to be due to either very small deletions or changes in the α_2 gene.

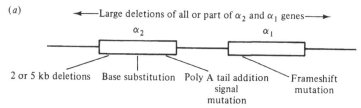

Fig. 7.23. Molecular defects in the α and β thalassaemias. The thalassaemias are a group of disorders characterized by an imbalance of globin chain production and are named as α and β thalassaemias according to which globin chain is affected. In some cases no α chains are produced (α^0 thalassaemias) in others defective α chains are produced (α^+ thalassaemias). The α^0 thalassaemias result from major deletion of portions of the α globin gene cluster, the α^+ thalassaemias can be due to small deletions or to any of the causes shown in the diagram which summerizes non-deletional α^+ thalassaemias. The β thalassaemias can also be β^0 or β^+ as above, but deletions are much less common.

The deletion of five bases (TGAGG) of the first intron immediately following the 5' invariant G, makes correct splicing impossible. A single base substitution (CTG→CCG) in the α_2 gene leads to leucine being replaced by proline, hence helix formation in the completed protein is disrupted and the α-chain is unstable. A single base change in position 14 of the α_2 gene leads to a new reading frame containing a stop codon (TGA) a few bases further along. A mutation in the α_2 chain AATAAA polyadenylation signal leads to defective poly A tail addition and mRNA instability. Several α-thalassaemias are known which are termination codon mutations, i.e. the UAA stop codon becomes AAA (lysine), GAA (glutamate), etc. Instead of chain termination, these amino acids are inserted and translation carries on for a further 30 or so amino acids producing a defective α-globin chain.

The molecular defects leading to the β-thalassaemias are similarly complex and about 30 different lesions are now known. Major gene deletions as with the α^0-thalassaemias are uncommon. Apart from chain termination mutations (e.g. UAU coding for tyrosine changing to UAA, a stop codon), frameshift mutations and poly A addition signal mutations as in the α^+-thalassaemias are also known. Mild forms of β^+-thalassaemias show mutations in the 5' 'upstream' regulating sequences TATA and CCAAT. Several 'splicing' lesions have also been defined in the β^+-thalassaemias, including the following:

1. Deletion of the invariant GT (donor) and AG (acceptor) site nucleotides at the intron/exon junctions of intron 1 or 2.
2. Single base changes in the (relatively) conserved sequence at the 5' end of intron 1. This leads to a reduction in normal splicing but surprisingly to the use of donor-like sequences in both exon 1 and intron 1.
3. Mutations leading to the generation of new splicing sites, e.g. a new acceptor site can appear in intron 1 leading to a non-functional mRNA containing some intron sequence.
4. The activation of 'cryptic' splicing sites within exon sequences. The mutation which gives rise to HbE also produces a new donor splice site within the exon 1 codon sequence. This is then used at a low level producing a defective mRNA lacking some exon sequence.

The genetic defects underlying faulty globin chain production have shown that what clinically appears to be a 'single-gene' disorder can be due to a lesion virtually anywhere in the transcription unit, i.e. within the coding portion, the intron portion or the flanking sequences. It is also significant that all of the defined lesions of globin chain synthesis lie close to or within the α- and β-globin gene clusters, i.e. lesions in distant 'regulatory' DNA sequences (if they exist) have yet to be defined.

Gene probes and restriction fragment length polymorphisms (RFLPs)

During meiosis, a single diploid pre-gametic cell (containing two copies of each chromosome, each with one copy of a particular gene upon it) divides into four haploid gametes (spermatozoa or oocytes), each containing one chromosome and one copy of each gene. At the first meiotic division, all the pairs of chromosomes (each one of which has already replicated into two sister chromatids) line up along their length and at this stage chromosomal material can exchange between sister chromatids belonging to different members of the chromosome pair (Fig. 7.24). This is known as 'crossing-over' or recombination. If two different genes are very close together on the same chromosome, they will not be separated by this event and will both finish up in the same gamete, i.e. they are said to be 'linked'. The further apart the two genes are on the same chromosome, the more likely they are to be separated and to finish up in different gametes. Thus, the frequency with which two genes (or loci) known to be on the same chromosome are separated into different gametes by recombination is a measure of how far apart the genes are on the chromosome. If two loci show such crossing-over in 1 % of gametes in a doubly heterozygous person, they are said to be one centimorgan (after T. H. Morgan, a pioneering American geneticist) apart. A centimorgan is estimated to represent 1000 kbp (i.e. a stretch of one million bp of chromosomal DNA) and the whole human genome represents approximately 33 Morgans (i.e. 3×10^9 bp).

Restriction endonucleases (p. 203) recognize specific sequences in double-stranded DNA and produce appropriate cuts. Some restriction endonucleases recognize short sequences and therefore produce frequent cuts in human DNA (e.g. Alu I–AGCT), some recognize longer sequences and therefore produce very infrequent cuts (e.g. Not I– GCGGCCGC). Assuming that enzyme digestion has gone to completion, cutting DNA with a particular restriction enzyme produces 'restriction' fragments of varying lengths depending on the distance apart of the sites which the enzyme recognizes. Variation in the position of each restriction site in the DNA of different individuals or between the DNA contained on homologous chromosomes will produce variation in the length of the restriction fragments after digestion, hence 'restriction fragment length polymorphisms' or RFLPs. Restriction sites are

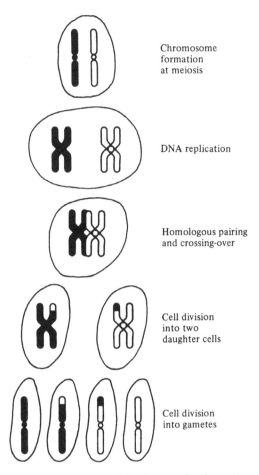

Chromosome
formation
at meiosis

DNA replication

Homologous pairing
and crossing-over

Cell division
into two
daughter cells

Cell division
into gametes

Fig. 7.24. The recombination mechanism. At meiosis, chromosomal material can be exchanged between homologous chromosomes. See text for details.

inherited in a stable Mendelian fashion and can therefore be regarded as genetic loci amenable to linkage studies as with defined genes. It has been estimated that approximately 400 such polymorphic loci, provided they were evenly spaced throughout the human genome, would be sufficient to give a reasonable chance of establishing linkage to the genes responsible for human disease states. Note that, so long as a disease is inherited, it is not necessary to have defined the genetic defect: the presence of the defective gene can be diagnosed by RFLP analysis provided that linkage to the polymorphic site is sufficiently close.

Major deletions of regions of DNA such as those seen in the α^0-

thalassaemias can easily be detected by restriction mapping of genomic DNA (Fig. 7.25). Single point mutations may also sometimes be directly detectable, a classic example being sickle-cell anaemia itself, where the base change GAG→GTG (i.e. a glutamate to a valine codon) occurs in a CCTGAGG sequence. This is a recognition site for Mst II and is

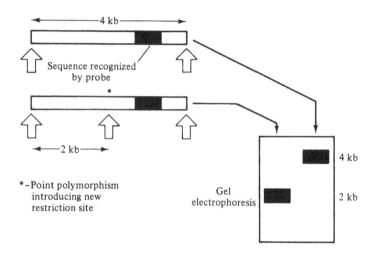

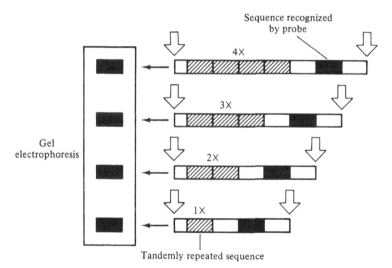

Fig. 7.25. Restriction fragment length polymorphisms. These can be due to a single base change which can introduce (or remove) a restriction enzyme site, hence using a probe for the shaded area shows differing sizes of DNA on digestion. Polymorphism can also be caused by insertion of variable length DNA segments between two fixed restriction enzyme sites.

abolished in sickle-cell β-globin which now reads CCTGTGG. This results in a larger molecular weight fragment (1.3 kb rather than 1.1 kb) on Mst II digestion of sickle-cell DNA. Most point mutations (possibly three-quarters) do not alter restriction sites and therefore are not detectable by restriction fragment analysis. If the sequence and the point mutation is known (i.e. if a large proportion of the defective gene has been sequenced and the variation from the normal gene is known), an alternative approach is to use oligonucleotide probes. Most 'gene probes' are several hundred (or even a few thousand) base pairs long and are usually prepared by cloning cDNAs or fragments of genomic DNA and then labelling by an appropriate procedure. Such probes hybridize over long stretches of DNA and are insensitive to single base pair mis-matches introduced by point mutations. Short oligonucleotides (say, 17–20 nucleotides long) can detect single base pair mis-matches. (Note that modern methods make possible the synthesis of defined sequence nucleotides several hundred bases long.) Genomic DNA is digested (using a suitable restriction enzyme or combination of restriction enzymes) to give fragments of around 1.0–1.5 kbp and then probed with radioactive oligonucleotides 17–20 bases long. Normal DNA and DNA with the point mutation are probed in parallel using one probe with a single base alteration complementary to the single base change produced by the point mutation. By choosing an appropriate temperature, the single base-pair mis-match will cause dissociation of the normal probe from the mutant DNA and vice versa. (The appropriate temperature for dissociation can be calculated roughly knowing the length of the probe, its G + C content, and how close the single base-pair mis-match is to the ends of the probe.) Oligonucleotide probes have successfully been used in the prenatal diagnosis of several forms of thalassaemia and also in ZZ α_1-antitrypsin deficiency, in each case the point mutation not being detectable by restriction enzyme analysis. Since clinically identical syndromes can be caused by point mutations at many different places in a transcription unit (e.g. the thalassaemias), several different oligonucleotide probes will be needed for each 'single-gene' disorder.

RFLPs in the diagnosis of human disease

'Anonymous' DNA probes are probes that have been selected randomly from genomic libraries on the basis that they recognize single-copy sequences on blots of digests of genomic DNA and not repetitive sequences. The region of DNA recognized by the probe is unknown and could be part of a structural gene or unrelated to a structural gene. The probe is then tested to see if it can detect RFLPs between unrelated individuals. In order to assign the RFLP locus to a particular chromosome, the genomic DNA from human–rodent somatic cell hybrids containing different human chromosomes is examined, or the probe is used directly on preparations of human chromosomes with *in situ* hybridization techniques (Fig. 7.26). Individual human chromosomes can be separated and sorted by fluorescence-activated flow sorting (p. 191) and genomic libraries constructed using DNA from each chromosome. 'Anonymous' DNA probes have been in use since approximately 1980. RFLPs arise either from point mutations affecting the recognition site for a particular restriction endonuclease (the vast majority of RFLPs) or from the insertion or deletion of DNA leading to a variation in the length of a restriction fragment. Several hundred 'anonymous' DNA probes have been catalogued by the Human Gene Mapping Workshops which are held every 2 years. Each probe is named beginning with 'D' (for DNA segment), and followed by the chromosome number to which the DNA segment maps. This is followed by 'S' (if it is a single copy sequence) or 'NF' if the probe hybridizes to numerous fragments. The final number in the locus name is that given sequentially by the Human Gene Mapping Workshop. Thus, the first anonymous DNA probe described detected locus D14S1, the first locus mapped to chromosome 14. Apart from catalogued probes, many other 'private' probes are in existence which have not been classified.

The application of RFLPs and cytogenetic methods can map a disease locus to within approximately several million base pairs. This approach, in which no knowledge of the protein product of the gene is available but the gene is mapped as a stretch of DNA within which mutations give rise to a specific disease, has been termed 'reverse genetics'. Two common diseases in which 'reverse genetics' have been applied are Duchenne muscular dystrophy and cystic fibrosis.

Duchenne muscular dystrophy (DMD) is an X-linked recessive disorder affecting approximately 1 in 4000 live male births. Affected boys show progressive muscle wasting (leading to very high levels of creatine kinase enzyme activity detectable in serum), lose the ability to walk by early adolescence, and die in their late teens or early twenties. One-third of cases are due to new mutations. Cytological studies have shown abnormalities in the short arm of the X chromosome at band 21 (Xp21). A probe defining locus DXS164 could be shown to map in this region and to be linked to the disease. Chromosome 'walking' bidirectionally from the locus defined by this probe produced further probes and eventually

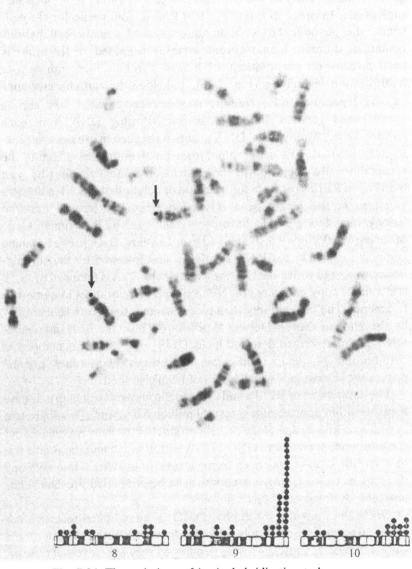

Fig. 7.26. The technique of *in situ* hybridization to human chromosomes. The example shown localizes a specific gene to one end of chromosome 9. The two 9 chromosomes are arrowed in the chromosome 'spread', the diagram of chromosome 9 (p is the short arm, q the long arm) shows the banding pattern and the accumulated score of photographic grains along the chromosome when nearly 100 dividing cells are surveyed. The grain distributions over chromosomes 8 and 10 are also shown diagrammatically. Figure courtesy of Dr S. Craig.

the whole area of the DNA in the Xp21 region covering the DMD gene could be mapped. Several of these probes are useful in prenatal diagnosis.

The gene which is defective in Duchenne muscular dystrophy is very large (2000 kbp) and is the largest human gene yet discovered. The protein product of this gene is coded for by over 60 separate exons and is 3685 amino acids (400 kDa) long. This large protein has been called 'dystrophin'. Its sequence shows homologies in different regions to both spectrin and α-actinin (see Chapter 2), and hence dystrophin appears to be a rod-shaped molecule possibly capable of binding to muscle membranes via its C-terminal region and to actin filaments via its N-terminal region. Dystrophin appears to be absent in muscle biopsies from patients with Duchenne muscular dystrophy and appears in normal muscle to be preferentially localized in muscle fibres specialized for fast contraction.

Cystic fibrosis is an autosomal recessive disease with a carrier rate of 1 in 20 among Caucasians and 1 in 2000 newborns are affected. The disease results in overproduction of mucus (which blocks airways and exocrine duct secretions and leads to chronically infected lungs), deficient pancreatic secretions and malabsorption. Epithelial cells show deficient chloride secretion and it is possible that the product of the cystic fibrosis gene is a protein which is part of a complex coupled to a chloride channel. The cystic fibrosis gene is known to be on chromosome 7 and close to a locus D7S8. By fragmenting human chromosomes and transfecting mouse cells, a small (10^6 bp) fragment of chromosome 7 has been isolated and shown to contain the D7S8 locus. By defining this fragment further, it may be possible to isolate the gene coding for the defective protein in cystic fibrosis.

Hypervariable regions in the human genome

These are regions of DNA sequence which vary in length between 'fixed' restriction enzyme sites. The most closely studied have been those associated with the insulin, α-globin and myoglobin genes. Approximately 360 bases upstream from the start of the insulin gene, a sequence of nucleotides (AGAGGGGTGTGGG) is repeated in tandem. Allelic variation in this region is due to differences in the number of times the copy of this sequence is repeated. The number of repeat copies can be grouped into three classes, one with 40 tandemly repeated copies, one with 95 and one with 170. Some of the polymorphisms in this region appear to be associated with the development of diabetes more often than can be accounted for by chance. Construction of a probe based on a tandemly repeated 33 bp sequence found in the first intron of the myoglobin gene has led to a new dimension in 'fingerprinting' individual human genomes. These short 'minisatellite' tandemly repeated sequences appear to be highly variable in the number of repeat copies, scattered throughout the genome (but probably not on the X and Y chromosomes), and inherited in a stable Mendelian fashion. Because this probe can pick up many highly variable loci simultaneously, the pattern of fragments following restriction enzyme digestion are totally specific for each individual. Such a polymorphism has important practical applications in forensic medicine, pedigree and paternity analysis and other familial genetic studies.

Genetic screening and gene therapy

It has been estimated that, of every 200 newborn babies, nine will go on to develop diseases such as diabetes, heart disease and schizophrenia, which are currently regarded as the end result of many different causes, some of which are genetically based (i.e. these diseases are 'multifactorial' in aetiology). Eight babies will have a significant congenital malformation (e.g. hare-lips, cleft palates, club feet), and three will have mental retardation of unknown cause (i.e. 'idiopathic' mental retardation). Two will have a detectable 'single-gene' disorder, and one will have a chromosome abnormality of which the commonest is Down's syndrome (trisomy 21). The purpose of 'genetic screening' programmes is to detect individuals who are at risk (or whose offspring are likely to be at risk) from genetically determined disease or genetically determined susceptibility to environmental agents.

Since it is clearly not possible to screen everybody for all diseases known to have a genetic cause, such programmes are necessarily highly selective. Decisions as to which diseases to screen for and in whom are dependent on many complex and sometimes controversial reasons. These include not only technical feasibility (i.e. is the underlying genetic cause understood and is a simple and reliable test available?) but also economic considerations, e.g. most single-gene disorders are so rare that it is difficult to justify the cost of screening many thousands of people to find a single case of the disease. Finally, genetic screening is particularly an area of medicine where ethical and legal aspects play a major role in the design and implementation of detection programmes.

Currently, genetic screening can be divided into screening of the neonate (newborn), prenatal screening and screening of potential carriers of genetic disease. Screening of the newborn has been carried out for many years and is centred on the detection of 'inborn errors of metabolism', i.e. an inherited deficiency of an enzyme in a metabolic pathway which leads to the accumulation of the substrate for the enzyme (or of metabolic products arising from utilization of the substrate by alternative secondary metabolic pathways). At least 200 such inborn errors are known and all inherited as recessive disorders. The commonest (1 per 10 000 births) is phenylketonuria (PKU) and this is routinely screened for in neonates in developed countries. The metabolic defect in PKU is shown in Fig. 7.27. The accumulation of high levels of phenylalanine in the blood (hyperphenylalaninaemia) can be detected by spotting blood on to filter paper and placing the paper in a bacterial growth inhibition assay (the Guthrie test). Hyperphenylalaninaemia leads to mental retardation by a mechanism which is not understood, but the following of a special diet low in phenylalanine allows normal mental development if begun in the first few weeks of life. The economic cost of caring for affected patients in institutions is far greater than the cost of the screening programme and prevention by dietary means.

PKU (Fig. 7.27) is the best example of effective neonatal screening for an inborn error of metabolism. Recently, screening for congenital hypo-thyroidism (cretinism) has been widely adopted. This can be carried out on the same filter paper as used for PKU screening and involves the measurement of thyroid-stimulating hormone (TSH). Although not an inborn error of metabolism in the classical sense, congenital hypothyroid-ism can affect about 1 in 4000–5000 neonates, i.e. a higher frequency than PKU. Effective treatment is available (thyroid hormone replacement) to arrest the mental retardation which would otherwise ensue.

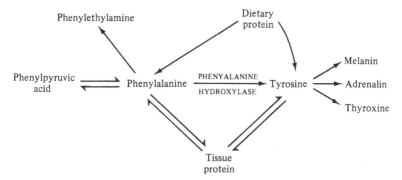

Fig. 7.27. The metabolic defect in phenylketonuria (PKU). In classical PKU, the enzyme phenylalanine hydroxylase is deficient, leading to an accumulation of phenylalanine in the bloodstream and abnormally high levels of phenylpyruvic acid. The disease is an autosomal recessive condition. Not all cases of raised blood phenylalanine are due to phenylalanine hydroxylase deficiency.

Prenatal (or foetal) screening most commonly involves examination of amniotic fluid, amniotic fluid cells or foetal tissue itself. The most common measurement carried out on amniotic fluid is the measurement of α-foetoprotein, which is produced by the liver as the foetal equivalent of serum albumin. In the presence of defective formation of the foetal spinal cord and brain (neural tube defects), increased levels of this protein are found in amniotic fluid and sometimes also in maternal serum. Amniotic fluid cells originate from foetal skin and can be grown in culture to produce sufficient material to examine cytogenetically or to measure specific enzyme levels. The commonest reason for amniotic fluid cell analysis is a maternal age of 35 years or greater leading to an increased risk of Down's syndrome (trisomy 21). Some of the inborn errors which have been diagnosed by enzyme measurements in amniotic fluid cells are shown in Table 7.4. Direct sampling of foetal blood by microtechniques is also possible and has been used to diagnose haemoglobin disorders and haemophilia. The disadvantage of amniotic fluid withdrawal (amniocentesis) and foetal blood sampling is that neither can be employed until relatively late in the pregnancy (3–6 months, i.e. the second trimester). If amniotic fluid cells have to be grown up further in tissue culture to provide enough material for chromosomal or enzymatic analysis or to obtain enough DNA for gene probe studies, then the firm diagnosis of a genetic defect may come even later in the pregnancy, possibly after foetal movements have been felt. Termination of pregnancy at this stage may therefore involve considerable maternal emotional trauma.

Early diagnosis (during the first 3 months of the pregnancy) is now possible by the use of chorionic villus sampling (CVS) which is explained in Fig. 7.28. Using this technique, it is possible to obtain enough foetal DNA to examine with suitable probes for restriction site changes, major deletions or rearrangements, RFLP linkages, or with oligonucleotide probes for single base mutations. While CVS still needs to be validated as a clinically safe procedure (current studies indicate an acceptably low foetal loss rate), the ability to diagnose genetic defects in the first trimester when termination is a much less traumatic procedure represents a major advance in prenatal diagnosis.

One problem in screening genomic DNA from individuals for defective genes is obtaining sufficient DNA for the probe to generate a detectable signal. The chorionic villus sampling method described above yields about 35 μg of DNA per sample. This compares with about 300 μg of DNA obtainable from the leucocytes contained in 10 ml of venous blood. While this is sufficient for several restriction enzyme digests, the need to be able to analyse smaller amounts of DNA does arise in some

Table 7.4 *Inborn errors of metabolism which have been detected by amniotic fluid measurements*
The major use of amniotic fluid measurements is the detection of α-foetoprotein which is present in raised amounts in neural tube defects (e.g. anencephaly and spina bifida). By culturing amniotic fluid cells (which are of foetal origin), the metabolic disorders shown have been diagnosed *in utero*. Data source as for Table 7.1.

Metabolic disorders	Defective enzyme or process
Lipid metabolism	
Fabry's disease	α-galactosidase A
Gaucher's disease	β-glucosidase
Gangliosidoses	β-galactosidases A, B or C
	Hexosaminidases A, or A and B
Krabbe's disease	Galactocerebroside β-galactosidase
Metachromic leucodystrophy	Arylsulphatase A
Mucolipidosis	Multiple lysosomal hydrolases
Niemann-Pick disease	Sphingomyelinase
Mucopolysaccharidoses	
Hurler syndrome	α-L-iduronidase
Hunter syndrome	Iduronic acid sulphatase
Sañfillipo A	Heparin sulphamidase
Maroteaux-Lamy	Arylsulphatase B
Carbohydrate metabolism	
Galactosaemia	Galactose 1-phosphate uridyl transferase
Glycogen storage II	α-1,4-glucosidase
Amino acid metabolism	
Arginosuccinic aciduria	Arginosuccinase
Homocystinuria	Cystathione synthetase
Maple syrup urine disease	Branched chain ketoacid decarboxylase
Methylmalonic acidaemia I	Methylmalonic-CoA mutase
Others	
Adenosine deaminase deficiency	Adenosine deaminase
Congenital erythropoietic porphyria	Uroporphyrinogen III co-synthetase
Lesch–Nyhan syndrome	Hypoxanthine-guanine-phosphoribosyl transferase
Xeroderma pigmentosa	Defective DNA repair
Hypophosphatasia	Alkaline phosphatases

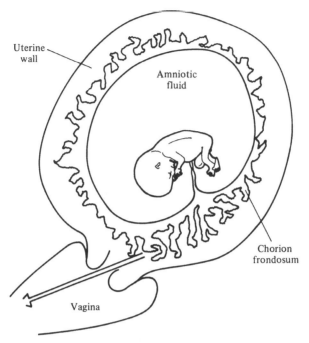

Fig. 7.28. The technique of chorionic villus sampling (CVS).
A major problem in prenatal diagnosis by recombinant DNA
methods is to obtain sufficient amounts of foetal DNA early in the
pregnancy. Up to about 11–12 weeks' gestation, DNA can be
obtained from the chorion (a foetal structure) as shown.

circumstances. The polymerase chain reaction (PCR) (Fig. 7.29) has
been designed to overcome this problem. Genomic DNA is heated to
separate double-stranded DNA. A pair of oligonucleotide primers is then
added, each of which is complementary to a sequence flanking the gene of
interest. DNA polymerase is added next. This extends each primer in a
$5' \rightarrow 3'$ direction, copying each separate strand of the original DNA
duplex. At the end of this synthesis, the reaction mixture is heated, the
two new double helices are thus separated into single strands and the
whole cycle is repeated many times with the number of copies of each
strand doubling with each cycle. The procedure can specifically amplify
the sequence of the gene of interest up to a million times and hence
massively increase the target for a probe directed against the gene
sequence. The use of a DNA polymerase which is heat-stable (from the
thermophilic bacterium *Thermus aquaticus*) and which therefore does
not need to be replaced at the beginning of each cycle has meant that the
polymerase chain reaction can be automated. This technique has been

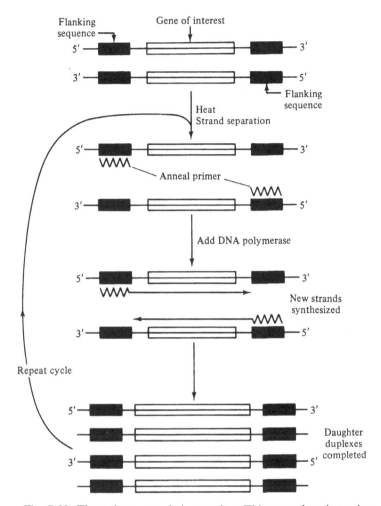

Fig. 7.29. The polymerase chain reaction. This procedure is used to amplify a particular sequence in genomic DNA by up to a million times, making it much easier to detect the sequence with a specific probe. Primers annealing to sequences flanking the gene are extended by DNA polymerase to produce two new complementary strands. After heating these strands are copied repeatedly in cycles of heating/synthesizing/heating. See text for further details.

applied in detecting rare HIV viral sequences in the DNA of infected patients before HIV-antibodies appear in the circulation, in the genetic screening of embryos produced by *in vitro* fertilization before implantation, and in analysing the DNA of epithelial cells produced by mouthwashes to determine cystic fibrosis carrier status.

The third area of genetic screening is represented by the detection of carriers of genetic (often recessive) disorders. Few inherited diseases occur sufficiently frequently in the general population to justify routine screening for the detection of carriers. However, some diseases are particularly common among particular racial, religious or ethnic groups, and screening of such a 'pre-selected' population may be justifiable. The best example of this is Tay-Sachs disease, the metabolic basis for which is shown in Fig. 7.30. This disease results in mental retardation, blindness, seizures, paralysis, and death in the first few years of life. Tay-Sachs disease is mainly seen among Ashkenazi Jews (incidence about 0.3 per 1000 births compared with 0.002 in all other racial groups) and measurement of hexosaminidase A levels can detect carrier status in adults and the presence of the disease in the foetus from examination of amniotic fluid cells. The incidence of Tay-Sachs disease has been significantly reduced by detection of carriers and genetic counselling.

The object of 'gene therapy' is to correct or ameliorate the clinical manifestations of a defective gene. Attempts to do this do not necessarily have to employ genetic techniques. For example, special diets can be employed in phenylketonuria and other inborn errors of metabolism, immunoglobulin can be given to children with genetically determined

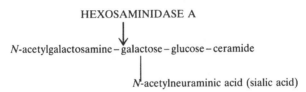

HEXOSAMINIDASE A

N-acetylgalactosamine – galactose – glucose – ceramide

N-acetylneuraminic acid (sialic acid)

Fig. 7.30. The metabolic defect in Tay-Sachs disease. Gangliosides are complex lipids containing sialic acid, sugar residues, and the amino-alcohol sphingosine. Gangliosides are found in the membranes of most cells but are particularly abundant in the brain (the name 'ganglioside' reflects their initial discovery in nerve ganglia). The structure shown is that of GM_2 which is one of the four major species of ganglioside. This structure is formed by the step-wise addition of the sugar residues shown to ceramide (N-acylsphingosine) and broken down by the step-wise removal of residues in the reverse order. The removal of N-acetylgalactosamine from GM_2 is catalysed by hexosaminidase A, an enzyme found in lysosomes. This enzyme is thought to have an $\alpha\beta_2$ trimeric structure and mutation at the locus on chromosome 15 coding for the α-subunit leads to hexosaminidase A deficiency (Tay–Sachs disease). Several other defects in ganglioside metabolism are known which lead to related gangliosidoses. Tay–Sachs is the first genetic disease in which the birth of an affected child has been prevented by mass screening of an at-risk population.

defects of the immune system, growth hormone (derived from human pituitaries or produced by recombinant DNA techniques) can be given in some forms of dwarfism, Factor VIII in haemophilia, and thalassaemic children kept alive by repeated blood transfusion. Some congenital abnormalities can be corrected fully by surgery. However, in general, 'replacement' therapy for genetic defects is not fully effective and is often expensive and uncomfortable for the patient. Furthermore, long-term replacement can create its own problems, e.g. the iron overload (partially correctable by chelating agents) seen following repeated blood-transfusions and the finite risk of transmitting hepatitis, AIDS or viral central nervous system disease by repeated infusion of human protein products. Attempts to replace missing enzymes in metabolic pathways by direct supply of the enzyme in liposomes or red cell ghosts (see Chapter 4) have been disappointing and the problems of 'targeting' the enzyme to the correct tissue(s) are complex. The above, coupled with the problems of preventing genetic disease by screening procedures outlined above, has led to the serious consideration of 'curing' genetic defects by introducing a correct version of the gene into the defective genome.

'Direct' gene therapy of this nature has still to be achieved in humans, although its feasibility has been quite clearly demonstrated in animals such as mice. Introduction of a foreign gene into the oocytes of such animals has been shown to lead to the expression of that gene in the adult animal, sometimes in a tissue-specific manner. Such animals are said to be 'transgenic'. With gene therapy of human genetic defects, it appears that such insertion of genes directly into germ-line cells is unacceptable to society and the current problem is how to insert new genetic material into somatic cells. Criteria for human diseases in which direct gene therapy can be attempted are that the cellular and molecular pathology of the disease is understood, that the appropriate normal gene has been cloned and that the gene is simply regulated, i.e. that it is always 'on'. Additionally, animal studies should show that the gene can be introduced into the correct target cell and that it can be suitably expressed therein, and furthermore that the introduction of the new gene does not harm the animal. In practice, gene replacement therapy in humans is currently restricted to those cells which can be removed from the body, treated *in vitro*, and returned to the same patient, i.e. cells from the bone-marrow and to a lesser extent the skin and liver. Three diseases which are potentially amenable to such gene replacement are glucocerebroside deficiency (Gaucher's disease), and adenosine deaminase and purine nucleoside phosphorylase immune deficiencies.

Introducing the new gene into the recipient cells *in vitro* can be attempted in several different ways. Calcium-phosphate mediated DNA uptake (p. 201) is too inefficient (1 in 10^{-10}–10^{-5} cells transfected) especially considering that the target stem cells constitute only 0.01 % of total bone-marrow cells and the most promising current technique is to use retroviruses to carry the required gene into the target cell genome. Several retroviral vectors have been designed for this purpose. In general, these vectors produce a high degree of transfection (up to 25 %) but, as yet, demonstration of expression of the gene *in vivo* in a clearly therapeutic way is lacking. Retroviral vectors integrate at random into the host genome and carry the risk of both inactivating normal genes or activating unwanted cellular genes. In particular, retroviral insertion carries the risk of activating cellular oncogenes. This risk can be minimized by removing promoter/enhancer sequences from the 3' LTR (p. 251) of the retrovirus by genetic engineering. It is likely that human clinical trials replacing gene defects in haemopoietic cells will begin in the near future.

Retroviruses and oncogenes

Cells growing in tissue culture normally require an adherent surface and a high level (say, 5 % or more) of serum included in the medium. Once the cells form a confluent layer, growth stops (contact inhibition). Infection of such cells by certain DNA or RNA viruses can result in the cells requiring much less serum, in a loss of contact inhibition so that the cells grow over each other, and in the acquisition of the ability to grow freely in suspension rather than requiring a flat adherent surface. Furthermore, such cells, when injected into suitable inbred animals, can lead to the production of solid tumours. This change in cellular behaviour is known as 'transformation' and the transformed cells have essentially acquired malignant characteristics. Transformation can be caused both by DNA viruses and by RNA viruses, i.e. by viruses which carry their genetic information in DNA or RNA viral genomes, respectively.

RNA tumour viruses are termed 'retroviruses' because their genetic information is copied from RNA into DNA by the enzyme reverse transcriptase, i.e. in the reverse direction from the normal flow of information from DNA to RNA. The earliest retrovirus studied was Rous sarcoma virus (RSV), which causes connective tissue cancers (sarcomas) in chickens. The RSV virion contains two copies of a 9.4 kb RNA, each beginning at the 5' end with a cap and the following sequence:

$$m \quad GpppGCCATTTTACCATT$$

This sequence is repeated at the 3' end of the molecule between a polyadenylation signal (AATAA) and a poly-A tail (Fig. 7.31). Between a 5' untranslated region (U5) and a 3' untranslated region (U3), there are four genes (spanning about 7.5 kbp) called *gag*, *pol*, *env* and *src*. The rest of the RSV virion consists of about 50 copies of reverse transcriptase protein, several thousand copies of four virion proteins (12–27 kDa), and the envelope protein covering the surface of the virion. Once the retrovirus enters the cell, the RNA is copied into DNA by viral reverse transcriptase to give an end product of a double-stranded DNA with the structure shown in Fig. 7.31. The exact mechanism by which this is achieved is complicated and involves 'jumps' of portions of the newly synthesized DNA from one end of the RNA template to the other. Each end of the double-stranded DNA has been enlarged (compared to the initial RNA template) and now contains long terminal repeats (LTRs) as shown in Fig. 7.32. This DNA enters the nucleus and circularizes, and is

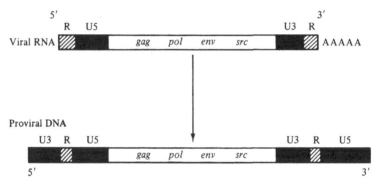

Fig. 7.31. The structure of the Rous sarcoma virus (RSV) genome. The RNA of the Rous sarcoma virus is 9.4 kb in size and contains four genes as shown. The transforming oncogene is *src* (for *sarc*oma). At each end of the genome there is a repeated sequence (marked R) of GCCATTTACCATT, which separates the untranslated 5' region (U5) from the 7-methylguanosine cap at the 5' end of the RNA and also separates the 3'-untranslated region (U3) from the poly-A tail. This RNA is copied into DNA by reverse transcriptase by a complicated procedure producing the proviral DNA as shown. The U3 sequence is moved to the 5' end and the U5 sequence to the 3' end, so that each end of the DNA now has a U3/R/U5 arrangement which is known as the long terminal repeat or LTR. The LTR contains the viral promoter and is essential for transcription of the viral genome once integrated into the host genome. The sites of insertion of the virus into host DNA appear to be random, but if they are inserted close to an oncogene, for instance, they can have profound effects on cellular physiology.

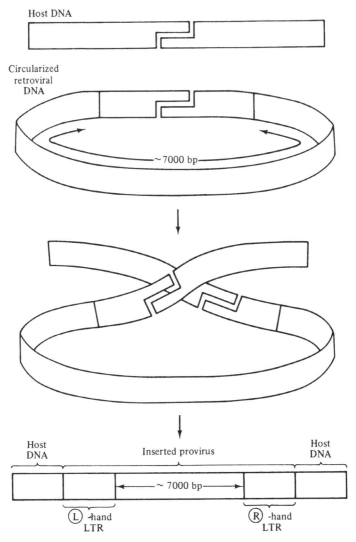

Fig. 7.32. Insertion of a retrovirus into host genomic DNA.
Circularized retrovirus DNA is inserted randomly into host DNA.
Here the host DNA is cut at a randomly occurring sequence
partially complementary to the circularized retroviral sequence
between the two original LTRs. Final 'tailoring' of the insert site
results in the inserted retrovirus (the provirus) containing 5′ and 3′
LTRs which are essential in transcription of the proviral genome.

then inserted into the host DNA. The site at which the viral DNA is cut is specific; the site of integration into the host DNA is random.

Once the DNA has been inserted, the host cell machinery is used to produce RNA transcripts which serve both as mRNAs for virion proteins (the *gag* gene for a large polypeptide which is proteolytically processed into four smaller virion proteins, the *pol* gene for reverse transcriptase, and the *env* gene for envelope protein) and as new viral RNA genomes. In this transcription, the LTR regions are critical in that the 5′ LTR contains promoter and enhancer activity and the 3′ LTR the polyadenylation signal (although how these identical sequences perform these different functions is not clear).

The infected cell continues to make new virus particles and to pass the parental DNA on to daughter cells. Retroviruses do not normally kill cells. If retroviral proviruses are inserted into the DNA of germ cells in the whole animal, the provirus is passed on from generation to generation, however such insertions are usually transcriptionally inactive. It has been estimated that 0.1–1.0 % of the mouse genome may, in fact, be composed of retroviral sequences.

None of the *gag*, *pol* or *env* gene products transforms cells and viral reproduction does not require transformation. This is brought about by the product of the *src* gene, which is sufficient in itself to transform fibroblasts in culture and endow them with the capacity to produce tumours in the whole animal. The retroviral *src* gene is therefore an oncogene (i.e. a tumour-causing gene). The retroviral *src* gene originates from host DNA, i.e. the retrovirus has, in the past, 'picked up' the *src* gene by integrating close to a normal *src* gene in the host DNA. This process is known as 'transduction' and the RSV is a transducing retrovirus. The viral form of the oncogene is referred to as *v-src*, the normal cellular counterpart (or proto-oncogene) as *c-src*. At least 20 different retroviral oncogenes and their cellular proto-oncogene partners are now known, and they are named according to a three-letter code word (e.g. *src*, *myc*, *fos*, *erb*, *abl*) and with a 'v' or a 'c' prefix to distinguish the viral or cellular version of the gene. It is assumed that the roles of the cellular proto-oncogenes (which appear to be highly conserved throughout evolution) lie in normal pathways of differentiation and cell growth which become deranged when the oncogene is transduced into its retroviral version. The conversion of a cellular proto-oncogene into a retroviral oncogene may just involve a quantitative increase in expression as the transcription of the gene occurs under the influence of the retroviral LTR promoter and enhancer elements, or it may involve an alteration of gene structure, which, in some cases, is known to be merely a single base mutation.

Cellular proto-oncogenes (and retroviral oncogenes) are known to code for specific proteins (Fig. 7.33). One class of proteins are tyrosine protein kinases (produced by *src*, *abl*, *erb-B*, etc.). These phosphorylate tyrosine residues in proteins (in contrast to the much commoner phosphorylation of serine and threonine residues) and many of this class of oncogene products are bound to the plasma membrane. Normal cell-surface receptors such as the epidermal growth factor (EGF) receptor,

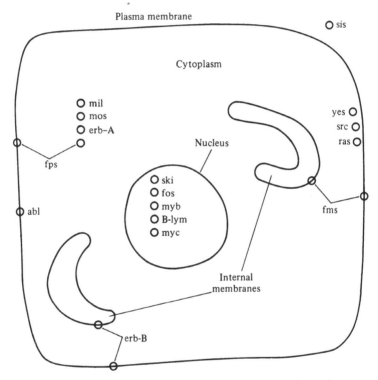

Fig. 7.33. Mechanisms of action of oncogenes and their products. Some oncogene products (e.g. *src*, *yes*, *abl*) are protein kinases which specifically phosphorylate tyrosine residues in susceptible proteins. Often (but not always) these oncogene products are bound to the plasma membrane. Other plasma membrane bound oncogene proteins (e.g. *ras*) are GTP-binding proteins. Cytoplasmic oncogene proteins (e.g. *mil*, *mos*) appear to resemble protein kinases while one oncogene product (*sis*) is related to platelet-derived growth factor. The other major group of oncogene proteins (e.g. *myc*, *myb*, *fos*) are found in the nucleus and are believed to regulate gene activity by binding to specific regions of DNA.

the platelet-derived growth factor (PDGF) receptor, and the insulin receptor are now known to consist, in part, of tyrosine protein kinases which phosphorylate specific tyrosine residues in the receptor following binding of the specific ligand. This suggests that tyrosine phosphorylation may be important in regulatory 'signalling' pathways, but exactly which proteins are important in tyrosine phosphorylation by oncogene products is not yet clear.

Another class of plasma membrane bound proteins are the 'p21' proteins exemplified by the Harvey murine sarcoma (*Ha-ras*) and Kirsten murine sarcoma (*Ki-ras*) oncogene products. These are GTP/GDP binding proteins with GTPase activity, as seen in the G-protein associated with adenylate cyclase (p. 138). In fact, G-proteins and *ras* gene products show some homology in sequence. A third group of oncogene products (*myc*, *fos*, *myb*) are nuclear proteins which are possibly involved in regulating transcription. Finally, some oncogene products (*sis*) appear to be closely related to normal growth factors (PDGF in this case). Thus, the retroviral oncogene products appear to act on pathways normally utilized for cell differentiation and responses to normal controlling factors. Retroviruses are also known to produce cancers in more slowly responding systems, e.g. some avian leukaemias. In these, the retrovirus appears to insert close to a normal cellular proto-oncogene and increase its transcription (leading to cancer) by providing either promoter or enhancer (or both) activity present in the LTRs.

Some oncogenes have been identified in human tumours which are not associated with retroviruses but which are mutated genes present in cellular DNA. Those which have been identified are related to the *c-ras* proto-oncogene, sequencing of such a cellular oncogene (from a bladder carcinoma) has shown that the difference between the normal cellular gene and the oncogene was just one base substitution. This was a G→T change, leading to a valine for a glycine at position 12 of the p21 protein *ras* gene product, a change which was sufficient to remove the GTPase activity of the *ras* protein. Other cellular oncogenes are known to be turned on by chromosomal translocations, e.g. chronic myeloid leukaemia in Man is associated with the Philadelphia chromosome which is the fusion of chromosome 9 to a piece of chromosome 22. The *c-abl* gene lies at the break-point on chromosome 9 and, on transcription, the resulting mRNA appears to be derived partially from *c-abl* and partially from a sequence contained on chromosome 22. Some human tumours (especially of nervous tissue) show duplication of the DNA of localized regions of chromosomes. Oncogenes, especially those related to *c-myc*, have been shown to be near these regions.

In summary, proto-oncogenes can be converted to oncogenes by retroviral transduction or by the adjacent insertion of retroviral DNA. Cellular oncogenes can be produced by point mutation, translocation or duplications. Although oncogenes can be demonstrated in human cancers, their causal role and its mechanism(s) remains to be established. One human cancer (human T-cell leukaemia) appears to be clearly associated with a retrovirus (HTLV). A related virus (HTLVIII) causes acquired immune deficiency syndrome (AIDS) with an associated rare skin cancer called Kaposi's sarcoma. Unusually for a retrovirus, HTLVIII kills the host T-cell, thus paralysing the immune system. How far oncogenes are responsible for more common human tumours such as lung, colon, breast, prostate and bladder cancers is not clear. It is currently felt, however, that while tumours may possibly be initiated by the activation of one oncogene, the development of the fully malignant state may require cooperation from several oncogenes. Furthermore, it is clear from studies on cancers in Man which can occur in either sporadic or familial forms (particularly tumours of the retina and kidney) that genes exist which can prevent the development of malignancy even if an oncogene is present. These are so-called 'anti-oncogenes'.

FURTHER READING

Alberts, B., Bray, D., Lewis, J., Raff, M., Roberts, K. and Watson, J. D. (1989). *Molecular Biology of the Cell*, 2nd edn. Garland: New York.

Darnell, J., Lodish, H. and Baltimore, D. (1986). *Molecular Cell Biology*. Scientific American Books: New York.

Finean, J. B. R., Coleman, R., and Michell, R. H. (1984). *Membranes and their Cellular Functions*, 3rd edn. Blackwell: Oxford.

Martin, B. R. (1987). *Metabolic Regulation: a molecular approach*. Blackwell: Oxford.

Stanbury, J. B., Wyngaarden, J. B., Fredrickson, D. S., Goldstein, J. L. and Brown, M. S. (1983). *The Metabolic Basis of Inherited Disease*, 5th edn. McGraw-Hill: New York.

Stryer, L. (1988). *Biochemistry*, 3rd edn. W. H. Freeman: New York.

Watson, J. D., Hopkins, N. H., Roberts, J. W., Steitz, J. A., and Weiner, A. M. (1987). *Molecular Biology of the Gene*, 4th edn. Benjamin–Cummings: Menlo Park, California.

Weatherall, D. J. (1985). *The New Genetics and Clinical Practice*, 2nd edn. Oxford University Press: Oxford.

INDEX

acquired immune deficiency sydrome, 255
actin, 59
actin-binding proteins, 59
actin genes, 231
active transport, 121
acute phase response, 26
adenine, 191
adenylate cyclase, 135
  activation of by cholera, toxin, 140
adipose tissue, 145
affinity, 41
affinity chromatography, 17, 43
AIDS, 255
alanine transaminase, 65
alcoholism, 62
alkaline phosphatase, 71
Alu sequences, 220
amino acids, 1–8
  abbreviations of, 5
ammonium sulphate precipitation, 13
amniocentesis, 244
antibodies, 29–52
  anti-idiotype, 48
  monoclonal, 38–52
  polyclonal, 29
antigenic determinant, 29
anti-oncogenes, 255
anti-protease, 7
α_1-antitrypsin, 23, 172
apoprotein B, 115
asialoglycoprotein receptor, 119
asparaginase, 73
Aspartate transaminase, 70
atherosclerosis, 115
autoimmune disorders, 230
automated DNA sequencing, 218
autophosphorylation, 159
autoradiography, 20, 169, 208, 215
autosomes, 188

avidity, 41

β cell, 173
B lymphocyte, 31
bacteriophage, 205
bacteriophage lambda, 210
Bence–Jones protein, 39

Ca^{2+}, intracellular, 154
caffeine, 136
calcitonin-gene-related-peptide, 227
calmodulin, 152, 155
cancer, oncogenes, 250
cap formation on mRNA, 224
cDNA, 207
cell fusion, 40, 137
central dogma, 194
centrifugation
  density gradient, 97
  differential, 96
  equilibrium, 97
  isopycnic, 97
centromere, 220
cholera toxin, 127, 140
cholesterol, 83, 114
chorionic villus sampling, 244
chromatin, 196
chromosome, 196
chromosome walking, 191
clathrin, 112
coated pits, 112
coated vesicle, 112, 172
coated viruses, 109, 114, 126
competitive binding assays, 35–8
complement, 127
coronary heart disease, 115
CRP, 27
C-peptide, 178
C-reactive protein, 27

creatine kinase, 70
crossing over, 235
CVS, 244
cyclic AMP, 135
    dependent protein kinase, 142
    in glycogen metabolism, 149–54
    and lipolysis, 145–8
    phosphodiesterase, 135
cyclic GMP, 154
cystic fibrosis, 126, 241
cytosine, 191
cytoskeleton, 56–63
cytotoxic therapy, 45

DAG, 155
defence
    cell-mediated, 31
    humoral, 30
desoxyribonucleic acid, 191
desmin, 60
desmosome, 105
diabetes mellitus, 185
diacylglycerol, 155
diagnostic imaging, 47
dibutyryl cyclic AMP, 136
diffusion
    facilitated, 121
    passive, 121
DMD, 239
DNA, 191
    cloning, 205
    library, 205
    repeated sequences in, 219
    repetitious, 219
    sequencing of, 211
    spacer, 219
    unclassified, 219
DNA ligase, 203, 207
DNA polymerase, 213
docking protein, 108, 171
dose response curve, 133
double helix, 192
Duchenne muscular dystrophy, 239
dystrophin, 241

ectoenzymes, 71, 90
electron microscopy, 91, 169
electrophoresis, 17, 20
ELISA, 73
emiocytosis, 166
EMIT, 73
emphysema, 23, 172
endocrine cell, 129

endocytosis, 110
    fluid phase, 111
    receptor mediated, 111
endosome, 112
enhancers, 162, 224
enzymes, 63–78
    as diagnostic reagents, 71
    as labels, 73
    as specific detoxifying agents, 76
    as therapeutic agents, 73
    as thrombolytic agents, 74
    clinically important, 68
    immobilised, 73
    tissue specific, 63–71
enzyme electrodes, 73
enzyme replacement, 77
enzyme-linked immunoadsorbent assay,
        73
enzyme-multiplied immunoassay
        techniques, 73
epitope, 29
euchromatin, 196
exocytosis, 166
exon shuffling, 12
exons, 223
expression vectors, 208

FACS, 45
fascia adherens, 105
fibrin, 75
fibronogen, 75
fluid mosaic model, 99
fluorescence-activated cell sorter, 45
fluorescence-activated flow sorting, 239
α-foetoprotein, 244
freeze–fracture, 95

G protein, 138, 255
ganglioside, 83
gap junctions, 105
gas–liquid chromatography, 83
Gel electrophoresis
    of DNA, 203, 211
    of proteins, 17, 21
gel filtration, 15
gene probes, 235
gene therapy, 242
genetic code, 191, 195
genetic disease, 189
genetic engineering, 205
genetic recombination, 235
genetic screening, 242
GLC, 83

glial fibrillary acidic protein, 60
globin genes, 232
gluconeogenesis, 227
glucose 6-phosphatase, 149
glucose tolerance test, 186
glucose uptake, 158
glycogen, 149
glycogenolysis, 149
glycogen phosphorylase, 150
glycogen synthase, 149, 153
glycoproteins, 11, 90
glycosylation, 109
Golgi apparatus, 169
Graves' disease, 165
guanine, 191

haploid, 235
hapten, 30
HAT medium, 41
HDL, 116
helix
  α, 3
  DNA double, 192
heterochromatin, 196
heterogeneous nuclear RNA, 194
HGPRT, 41
high-density lipoprotein, 116
histone, 194
hnRNA, 194
homogenization, 96
hormone
  action, 128–65
  definition, 128
  receptor, 133, 137, 159, 161
hybridization in situ, 208
hydrogen bonds
  in DNA, 193
  in proteins, 3, 8
hypercholesterolaemia, 118
hyperthyroidism, 164
hypothyroidism, 164, 243
hypervariable regions in DNA, 241
hypoxanthine guanine phosphoribosyl
    transferase, 41

IgA, 29
  pIgA, 119
IgD, 29
IgE, 29
IgG, 29
IgM, 29
immune
  response, 31

system, 30
immunoassays, 32–8, 41–3, 73
immunoglobulins, 29–52
  heavy chain, 29
  light chain, 29
  structure of, 29–30
immunoblotting, 18, 87
immunohistochemistry, 54–5
immunoprecipitation techniques, 32
immunoradiometric assay, 38
inborn errors of metabolism, 77, 188, 242
induction, 227
inositol triphosphate, 155
insulin, 176
  in diabetes, 185
  receptor, 159, 255
  secretion of, 179
intermediate filaments, 60–3, 199
introns, 12, 223
ion-exchange chromatography, 16
IP_3, 155
IRMA, 38
islets of Langerhans, 173
isoelectric focussing, 16
isoelectric point, 16
isoenzymes, 65–71

keratins, 60
Klenow fragment, 215

lactate dehydrogenase, 66
lamin, 199
LCAT, 115
LDL, 115
lecithin cholesterol acetyl transferase, 115
lipid, bilayer, 80
lipolysis, 145
long terminal repeats, 221, 252
low-density lipoprotein, 114
LTRs, 222, 252
lymphocytes (B,T), 31
lysogeny, 210

M13 phage, 215
macula adherens, 105
magic bullets, 45
Mallory bodies, 62
mannose 6-phosphate receptor, 173
MAPS, 58
Maxam–Gilbert DNA sequencing, 211
meiosis, 235
membranes, 79–107
  asymmetry of, 84, 90

membranes (*cont.*)
  bilayer structure of, 91
  lateral diffusion in, 100
  preparative isolation of, 95
  solubilization of, 87
  transport across, 121
messenger RNA, 192
methyl xanthine, 136
microfilaments, 59
microsomes, 96, 170
microtubule-associated proteins, 58
microtubules, 56–9
molecular sieving, 15
monoclonal antibodies, 38–52
mRNA, 192
myeloma, 39
myosin, 59
myxoedema, 165

N protein, 138
Na$^+$ K$^+$ ATPase, 124
Na$^+$ K$^+$ pump, 124
new genetics, 191
Northern blotting, 19
nuclear envelope, 199
nuclear lamina, 199
nucleosome, 196

oncogenes, 250
  encoding G protein homologue, 255
  encoding tyrosine kinase, 254
ouabain, 125

PAGE, 17
paraprotein, 39
patch-clamp technique, 184
PCR, 246
peptide bond, 2
pertussis toxin, 141
phagocytosis, 111
phenylketonuria, 243
phosphatidyl inositol, 4, 5
  biphosphate, 155
phospholipids, 82
phosphorylase, 150
phosphorylase kinase, 152
pinocytosis, 111
PIP$_2$, 155
PKU, 243
plasma membrane, 79–107
  bilayer structure of, 91
  disease and, 125
  domains in, 102

endocytosis and recycling of, 110
  fluid mosaic model of, 99
  interaction with cytoskeleton, 102
  lateral diffusion in, 100
  lipid asymmetry in, 84
  lipid bilayer in, 80
  preparative isolation of, 95
  proteins of, 86
  protein asymmetry in, 90
  synthesis and turnover of, 105
  transport across, 121
plasmids, 205
  pBR322, 207
plasmin, 74
plasminogen, 74
pleated sheet, β, 3
poly A tail, 225
polyacrylamide gel electrophoresis, 17
polyclonal antibodies, 29
polymerase chain reaction, 246
polymeric immunoglobulin A, 119
polysome, 169
post-translational modification, 11
primary structure, 2
primer in DNA synthesis, 213
proinsulin, 176
pro-opiomelanocortin, 227
protein kinase, C, 155
protein phosphatases, 153
proteins
  domains of, 9
  fibrous, 2
  folding of, 2–11
  globular, 2
  glycosylation of, 109
  hydrogen-bonding of, 3, 8
  peptide bonds in, 2
  purification of, 12–17
  secretion of, 169
  structure of, 1–12; primary, 2;
    quarternary, 11; secondary, 7;
    tertiary, 8
  serum, 20
  tissue-specific, 53–78
proto-oncogene, 253
pseudogenes, 219, 231
pulsed-field electrophoresis, 191
purine, 191
pyramidine, 191

quaternary structure, 11

radioimmunoassay, 35

ras proteins, 255
rate-limiting step, 143
receptor-mediated endocytosis, 111
receptor occupancy, 137
recombination, 235
respiratory quotient, 145
restriction endonucleases (restriction
    enzymes), 203
restriction fragment length polymorphism,
    235
retroviruses, 250
    AIDS virus, 255
reverse transcriptase, 194
RIA, 35
ribonucleic acid, 191
ribosomal RNA, 192
ricin, 45
RNA, 191
RNA polymerases, 192, 222
RNA synthesis, 222
rough endoplasmic reticulum, 108, 169
Rous sarcoma virus, 250
RQ, 145
rRNA, 192
RSV, 250

Sanger didesoxy DNA sequencing, 211
Scatchard analysis, 134
scanning electron microscopy, 95
screening DNA library, 208
SDS, 17
SDS-PAGE, 18, 87
SDS-polyacrylamide gel electrophoresis,
    18, 87
second messenger, 135
secondary structure, 7
secretagogues, 180
secretion, 166–87
    constitutive, 167
    triggered, 167
secretory component, 120
secretory vesicles (granules), 166, 171
SEM, 95
sequencing of DNA, 211
serum antiprotease, 74
serum proteins, 20
    albumin, 20
    globulin, 20
sickle-cell haemoglobin, 233
signal amplification, 142
signal hypothesis, 107, 170
signal peptidase, 108, 170
signal recognition particle, 108, 171

signal sequence, 108, 170
single-gene diseases, 188
single-stranded phages, 215
SLE, 230
small nuclear ribonucleoprotein particles,
    snRNPs, 230
sodium dodecyl sulphate, 17, 87
sodium-potassium pump, *see* Na$^+$ K$^+$
    pump
sorting of proteins, 172
Southern blotting, 19
spare receptors, 137
spectrin, 104, 241
splicing of RNA, 228
SRP, 171
steroid hormone action, 161–3
stimulus-secretion coupling, 176
structural genes, 219
subcellular fractionation, 95
superhelicity, 199
systemic lupus erythematosus, 230

T3, 163
T4, 163
T-lymphocytes, 31
TATA box, 224
Tay-Sachs disease, 248
telomere, 220
TEM, 91
tertiary structure, 8
thalassaemias, 233
thin layer chromatography, 83
thrombin, 75
thrombosis, 74
thymine, 191
thyroid disease, 163
thyroid hormones, 163
thyrotropin (thyroid stimulating hormone,
    TSH), 164, 243
tight junctions, 105
time course, hormone action, 133
tissue factor, 254–5
tissue-specific
    enzymes, 63–71
    proteins, 53–78
TLC, 83
topoisomerases, 199
TPA (tissue plasminogen activator), 75
transminase
    alanine, 65
    aspartate, 70
transcription, 192, 208, 222
transcription units, 232

transcytosis, 118
transfer RNA, 192
transformation by DNA, 201
transformation to cancerous state, 250
translation, 108, 192, 208
transmission electron microscopy, 91
triacylglycerol, 145
triplet code, 194
tRNA, 192
troponin C, 152
TSH, 164, 243
tubulin, 56
tubulin genes, 231
two-dimensional electrophoresis, 18

ultracentrifugation, 96
upstream elements, 162, 224
uracil, 191

vectors for cloning, 205
very low-density lipoprotein, 115
vimentin, 60
VLDL, 115

Western blotting, 18, 87

zonula adherens, 105